AUDREY THOMAS

Intertidal Life

Books by Audrey Thomas

Ten Green Bottles

Mrs. Blood

Munchmeyer and *Prospero on the Island*

Songs My Mother Taught Me

Blown Figures

Ladies and Escorts

Latakia

Real Mothers

Two in the Bush and Other Stories

AUDREY THOMAS

Intertidal Life

A NOVEL

Beaufort Books

New York Toronto

LIBRARY OF CONGRESS CATALOGUING IN PUBLICATION DATA

Thomas, Audrey Callahan.
Intertidal life.

I. Title.
PS3570.H5615 1984 813'.54 84-11158
ISBN 0-8253-0247-1

Published in the United States by
Beaufort Books
9 East 40th Street
New York, New York
10016

Published simultaneously in Canada by
Stoddart Publishing
A division of General Publishing Co. Limited
30 Lesmill Road, Toronto
Ontario, M3B 2T6

COVER DESIGN:
Brant Cowie/Artplus

COVER PAINTING:
Taylor Bay, Gabriola Island, 1952, oil on canvas, 61 × 76.2 cm, by E. J. Hughes
Photo: Tod Greenaway
Collection of Dr. Max Stern, Dominion Gallery, Montreal
Courtesy: Surrey Art Gallery, British Columbia

To Jonathan

Section I

''Part of the following day was employed in arranging and setting in order our records of observations, charts and calculations, and the notes made on all matters which having been jotted down on board ship in the midst of toil and labour required to be systematically expanded in order that they might be in good order and not convey a confused idea of the information gained. [We also replenished our supply of water, twenty barrels of which could be filled in a day in that district].''

—*A Spanish Voyage to Vancouver*

She was standing at the bottom of the government wharf and it was very early in the morning. The store above the wharf was not yet open and there were no cars on the road hurrying to catch the 8:00 A.M. ferry. Where had the fog come from? Last night had been so clear. After supper they had sat on the porch, talking, watching a flamboyant sunset (one expected trumpets at any moment) and then the dark had come slowly, softly, a calm exhalation. Finally the stars and a moon so clear it seemed transparent. But then in the middle of the night the foghorn began, un-hunh/uh-hunh, and Alice had sat straight up in bed, frightening the cat, who was lying across her feet. And had not been able to get back to sleep, too many dark thoughts and the foghorn out there, with its regular, monotonous lament, the lowing of drowning cattle, always, to her, a sound of unrelieved loneliness, irretrievable loss.

Once it began to be light she felt better, got up quietly and made herself a pot of tea, took it back to bed with her, wondering how soon the fog would lift, whether this would spoil their outing. Surely not; it was summer, after all. The sun would burn the fog away in no time. She had finished her tea and, wrapped in an old kimono, her daughter asleep in the other room, had gone down the path, crossed the road and walked to the very bottom of the wharf. The old cat, who had remained up at the top, complained loudly that she had not yet been fed.

This fog was so thick that only the slight sound of water lapping against the pilings indicated that the channel in front of her was still there. The other islands, the small ones nearby, the larger beyond and the distant mountains of Vancouver Island — all had disappeared. Alice had the feeling if she took one more step she might walk right off the end of the earth. Perhaps that wasn't the foghorn after all but the bellowing of those fabulous sea beasts who used to decorate the corners of old maps. What if this white curtain parted now, revealing dragons covered in glittering green scales, thick golden ropes between their teeth, towing the island away to — where? What a strange white swaddled world it was this morning, everything covered in a blanket of thick white wool, soft and wet and smelling strongly of the sea.

Years ago, when they had first come here, a pod of whales chose to swim through this channel on a moonless night. Peter had heard them first, for he had been outside on the porch smoking his pipe. He had come in and told her and the children, who were in bed but not asleep, and they had all trooped down to the wharf where they stood in the darkness while the great beasts puffed and spouted. All around them in the dark came the strange snorting noises of the whales. "I feel," Alice said to Peter later, curled up in bed, "like the first man might have felt, thinking all the dinosaurs were dead and then hearing one last lost tribe of them come snorting from the sea." The next morning, when he went out fishing, the whales came up all around him. He said it had been hard to sit still, imagined a smooth black back pushing up from underneath the boat, while he, poor tailless, finless brother was tipped playfully into the water. (Which was very cold on this side of the island. It made your bones ache even close to shore. The old-timers gave you half an hour in that water before you drowned. Peter was not wearing a life jacket, of course; he never did.) Later she took Flora, the youngest, almost four, on a plane trip to Toronto. As they flew over the Rockies the child would not sit still, kept shifting from one side of her seat to the other. Alice asked what was the matter, did she have to pee? "No,"

Flora said, distressed, "the mountains is comin' up through the sky." Alice remembered Peter and the whales and told him. He laughed. "I know exactly how she felt."

She went back up the steps, which were very steep because the tide was high, and then along the upper ramp. Spiders had built their webs between the red railings and in the middle of one of them, thrumming away, sat a very fat black spider, waiting for breakfast. The cat, seeing Alice, began to complain more loudly. "Shut up," Alice said, "all you think about is your belly and a warm place to sleep." She crossed the road and started up the path. The cat kept up her complaint. She was very smart and knew how to get around Alice. Three years ago she had become pregnant for the seventh time. Alice was fed up, tired of tacking notices to the bulletin boards at the south end, "Wanted: a home for adorable tabby kittens." "I am very cross with you," she told the cat. "The island is overrun with your kittens. This lot get drowned." The cat listened to her without blinking, giving off just the merest hint of a purr, then walked away. A few weeks later she went missing for a few days and returned much slimmer. But with no kittens—something had gone wrong. Alice, knowing it was stupid, couldn't help feeling a little guilty. Then the neighbor's child came up and asked if she could have one of Tabby's kittens. Alice was in the kitchen making soup. "I'm sorry dear," she said, "Tabitha never had her kittens, something went wrong."

The child stared at her. "I mean the ones under the house." And there they were, the usual five, a week old and eyes open. Tabby came around the corner with another one. "I can't drown them now," Alice said. Could she have, ever? Stones in a sack and then take the rowboat out, fling them into the middle of the channel? What culture was it where they used to put an adulterous woman into a sack with a cat and throw her in the sea? Couldn't remember. Her mind was full of bits of esoterica. But often not complete. Like knowing a phone number but not the person to whom it belonged. She had seen that thing on the adulterous woman in a book on the history of the cat. What book? Where?

That was the last litter, however. No more kitties under the house for Tabitha. Did she miss it? Nipples hardly even pinheads now. What was it like to be in heat only at certain times? Would it make life simpler or more difficult?

And there was Flora, *her* last kitten, at the top of the path in her white nightgown. How she had grown since last summer!

''Well, are we going?''

''Of course,'' said Alice. ''Just as soon as we are ready.''

''Are we going to take the car?''

''We are not. We are going to walk. It's much nicer going that way. And by the time we get there I expect the fog will have disappeared.''

Which it had, just as predicted, and the weather was once more behaving itself, the sky stretched tight and unfaded above them like a big blue sail full of wind. In the space of three hours not a single stray wisp of fog or cloud remained and the foghorn, thank heavens (for they were that much closer to the lighthouse now) had been turned off. Alice and Flora, each carrying a small rucksack, had walked almost to the northernmost tip of the island, a distance of only about half a mile, pausing just before the marina and skirting a small bay, crossing over to a catwalk which went up and around the cliff. There had still been fog then, although it had lightened somewhat, and looking down on the fish boats tied up below was like looking at the world through a thin gauze bandage. By the time they reached the top of the catwalk they were only about a hundred yards from the lighthouse, which looked out over the pass. (Porlier Pass, Boca da Porlier, named almost two centuries ago by a Spanish sailor.) The lighthouse keeper's wife had killed herself some years ago. She had been Danish or Norwegian and had committed suicide while home on holiday. Alice used to wonder whether she had not been able to stand the awful sound of the foghorn or whether she hated that regular, relentless sweep of light. Perhaps she couldn't stand islands.

After her death the lighthouse keeper had invited Alice and a friend to tea one afternoon. He showed them how everything worked (no more trimming of wicks or polishing the chimneys of the lanterns) and then they went inside the house and drank tea. This was not his first lighthouse; he produced his photo albums. Alice could not deal with his determined cheerfulness. Every time he smiled at them it was as though she heard, just for an instant, a small child sobbing in another room. Later her friend had said the visit made her "uneasy." "Yes," Alice said. She did not want to deal with anyone else's pain. Now he was married again and happy. The place looked very pretty seen through a layer of very fine gauze. There were beds of zinnias against the front wall and along with the smell of the sea came the smell of new-cut grass.

"Have you ever been there?" Flora asked.

"Once, years ago. That's what I was just thinking about. Why, would you like to go?"

It would be all right now, he had married again. It would be safe. The new wife kept a couple of small hairless dogs who were always trembling.

"All right," she said, "when I come back we'll arrange it."

But today they had turned away from the lighthouse and onto a path through the woods. A sign made from the bottom of a cardboard box was nailed to a tree. "Coon Bay, ½ mile." Alice was tempted to tear it down. Once in among the trees it was still quite cool and dark. "The first time we came here," Alice said, "your father had to carry you most of the way on his shoulders and so you were the first to see the water on the Gulf side. You clapped your hands."

That day the sun had been very hot so they built a lean-to from sticks and an old blanket. When Flora got sleepy they laid her down for a nap. And when she woke up they all took turns swinging her out and around and dipping her into the sea. The dog stood on the beach, barking excitedly until finally he could stand it no longer and plunged in himself. It was all different now. The little shacks were gone, the dog was dead. He had bitten a child who stuck its finger in his eye. This worried Alice so much she finally took him to the SPCA in Victoria and had him "put down," horrible phrase, but kept his collar still in the back of a bureau drawer. The hippies who had squatted there in the late sixties and early seventies had gone on to "fresh woods and pastures new." Goa. Matala. Who knows where. That first afternoon Peter and Alice had not really understood it all. It was a strange scene as they came up from the first bay to the second, through the cool, dark woods. There seemed to be a regular settlement of small shacks, some covered with cedar shakes, some merely wrapped in tar paper. And they were in various stages of repair. Here was one with a full set of windows, geraniums in juice cans and a bench beside the door. Here another with broken windowpanes and a rotting mouse-eaten mattress seen through a smashed-in door. A woman sitting on a stoop said she and her husband had been coming there for thirty-three years. ("We used to just fish with hand lines out there," she said, "the fish were always bitin'.") Others had been coming for longer than that, over from Chemainus and Ladysmith, workers in the big mills owned by the lumber company who also owned this land, Coon Bay. Allowed to build small shacks so long as they maintained the place. "This place even had a mayor once," the woman said, "it was real nice. Now those hippies have arrived. Always left everything unlocked. You can't do that any more. Can't even be sure if you do lock things up. This year all our mattresses had been slept on and peed on and God knows what else. And our ax was gone."

Peter and Alice smiled politely and led the children down to the water. Soft soft sand, biscuit-colored, and on one side a small lagoon, the water not warm but not that icy grip, that feeling of ice-cold glass, that you got on the other side. On one side of the bay the rocks stretched out in long humps and fingers, marvelous rocks, wrinkled and gray like the skin of old elephants but pocked and licked into fantastic shapes by the force of the winter waves. By the middle of the afternoon, in summer, these rocks became very warm and were lovely to lie upon, facedown and daydreaming. The warm rocks like the warm rough hide of friendly elephants. Even Alice could lie facedown, now, after all these years. ''You can always tell the mothers,'' Alice said once to a friend. ''They are the ones who are sitting up and not reading, staring out toward the water, ready to jump up and run.''

But still enough of a mother to realize how hot the sun was (for they had been here a while, had swum twice and dried off and eaten most of the picnic) and to sit up saying, ''Flora, will you please put on your hat?''

Flora did not answer but Alice knew she was not asleep. She looked at the backs of her daughter's legs, her smooth back and arms. How beautiful young girls were, young boys too, she supposed, although having no sons, she had no opportunity to observe them as closely as she had her daughters. Flora's skin was like the surface of a clear, unrippled stream. Please God, she said, I would like to see her grow up. And the other two, not quite grown yet, not quite flown away, not really. Please God. (She did not add, as she sometimes did when flying or in an automobile in awful weather, ''I'll never be bad again if you'll let me get out of this one.'' It was too serious this time. There was no god, of course, but fate, like the proverbial little pitcher, might have ears.)

''I wonder who first called us 'white,' '' Alice said, looking with pleasure at her daughter's skin, ''since what we really are is pink with a touch of yellow.''

Flora sat up, blinking against the sunlight. ''Flora, put on your hat,'' her mother said, automatically.

''It must have been someone dark,'' Flora said, ''so that we *looked* white to him. Or them.'' She pushed up the bottom of her bathing suit to examine her tan. ''Would you hand me my book?'' she said.

''Would you please please please put on your hat? I had sunstroke once and it wasn't very funny.''

''I know. You were out all day on the lake with your friend and when you sat up to dinner you both keeled over in a faint. My skin is much darker than yours,'' she said calmly. ''Much much much. You really are sort of pink. Pink with lots of freckles. Millions. I don't have a single freckle, not one.''

''Personally, I like freckles,'' Alice said, ''although my sister used to try and bleach hers with calamine lotion.''

''Well at least you didn't get that awful orange hair and really pale skin. At least you aren't one of those people who can't go out in the sun at all.''

''Even so, I can't see the point of staying out in the sun all day, doing nothing. I'm going to move in a minute, farther up under that tree.''

''I'll bet you lolled about in the sun a lot when you were my age.''

''Lolled generally, spent a lot of time in hammocks. But my sister was the real sun worshipper. I remember one summer, we had been to New York City and she had bought a two-piece, not so skimpy as a bikini but pretty daring, and there were ties you could pull to make it even more so. She spent practically every daylight hour laying on the beach on an old blanket, slathered in Johnson's baby oil.'' Alice thought of something.

''Lie back down for a minute,'' she said, ''on your stomach again.''

''Why?''

''I just remembered a game we used to play. We'd write on each other's backs and try and guess the words.''

''I thought you wanted me to put on my hat.''

''That too. But just lie back down.''

Alice leaned forward and began writing with her fingertip on the warm brown back. She wrote S-E-A.

''That's easy,'' Flora said, ''sea.''

''What's this?'' She wrote very fast.

''That's too fast. Do it again.''

''Nope,'' Alice said. She decided to change the subject. ''Another thing we used to do, or did one summer, was tape our boyfriends' initials on our backs with adhesive tape and let the rest of our backs tan naturally. After a couple of weeks

— zip — off came the tape and there we were, permanently branded. Well — temporarily.''

''What did your mother say?''

''She wasn't there that summer. It was just my grandfather and the housekeeper and us.''

''Where were your boyfriends?''

''They were counselors at a camp quite near. We met them at a square dance.''

''Was yours cute?''

''My sister's was cuter. Mine was nicer, I think.''

''Oh *nice,*'' Flora said. She sat up again and fished in her pack for the pocket mirror.

''My lips are too puffy,'' she said.

Alice laughed. ''That will be called 'kissy' in a couple more years.''

It was amazing to think that Flora had all that before her yet. Or did she? Alice remembered fumbling searchings, mouths like blind puppies, nuzzling, at some of the birthday and Halloween parties. Spin the bottle and post office, where, after the chaperoning mothers had gone upstairs for tea and chat, a boy would stand in a cupboard or behind a doorway and call out names. Once a boy whose face she could no longer remember — how was that possible? — summoned her and in the warm darkness leaned down and kissed her breasts, or the layers of cotton, then wool, above them. ''Postage due,'' he said, ''two cents.'' And laughed. Or barked. She felt as though she'd been stung by bees and went home right away without her coat, without the prize she won bobbing for apples, without saying thank-you very much I had a lovely time. She told her mother she was sick to her stomach and to prove it went in the bathroom, locked the door and stuck her fingers down her throat. What did that boy grow up to be? She hadn't been back to her hometown in years. She had no idea what had happened to any of them.

''Will you hand me my book,'' said Flora.

''Which one? *Madame Bovary* or *Nurse Prue in Ceylon*? They'll have to change that one to *Nurse Sue in Sri Lanka*.''

"Why?"

"Your knowledge of geography appalls me. Don't they teach it in school anymore? Sri Lanka is what used to be know as Ceylon. It's an island off the southeast tip of India. I imagine the guy in that one will be a tea planter."

"Well, that's the one I want."

"You sure you wouldn't rather have *Madame Bovary*?"

"Not just now." Flora put on her sun hat and Alice handed over the book, shaking her head and laughing.

"Oh yes," Flora said, "you read them too. You can shake your head all you like."

It had begun as a joke, reading Harlequins, and was still more or less a joke. At the beginning of the summer they had made up an elaborate list of things they intended to do (learn the names of the constellations, study seaweeds and intertidal creatures, paint the kitchen yellow) but a strange lethargy had overtaken them. It was partly due to the heat and partly due to a strong sense they had that they were waiting, that they could not go on until— Until. This had not been spoken about, or not in words, but sometimes Alice saw Flora looking at her very carefully. So long as the weather held, so long as the sea was there and the sunshine, then certain conversations would not take place. Perhaps, however, they should, thought Alice, gathering up her things, making ready to move into the shade. But why. She had gone over to see Peter, at the beginning, and had stayed the night. Downstairs on the couch in the two-car garage that was serving as a house until he got his real house built. She had not been able to sleep all night and when he "woke" her before dawn (she had a long way to go to catch the morning ferry and the unfamiliar road twisted and turned) she had started to cry. And then told him.

"That must be very worrying for you," he said. She knew he was concerned but he sounded very distant, like a psychiatrist or social worker. She told him she had not been able to sleep. She cried harder while he patted her in a nice brotherly fashion.

''You should have woken me up,'' Peter said. ''It wasn't good for you to lie here all night thinking dark thoughts.'' Alice imagined herself going up the ladder to the loft where Peter lay sleeping. He would have thought— What else could he have thought? And would he have taken her into his bed and held her, offered comfort? Or would he, before she had had a chance to whisper out her need — ''Peter, I'm frightened''—would he have said, half-asleep but wakened by her footsteps on the ladder, ''Now Alice, you know it's no good. Don't start anything, okay?'' But she hadn't even tried, she thought now, sitting on the warm rock and gathering her things together—she had simply lain there all night, below him, as though they were in separate bunks on a ship. He did not like hospitals. She had had her children alone; he would not even stay in the building. What made her think that now, when they weren't even married, he would come and hold her hand? She had been right not to beg to be comforted.

''What color would you call my eyes?'' Flora said.

''Blue. They're a nice blue.''

''*Nice*, again,'' Flora said. She took up the pocket mirror. ''You know what I'd call them? A 'just-blue' blue.''

''Does the man in this book have hawklike features?''

''Aquiline, this time. He has aquiline features. And dark hair, as usual.''

''Same thing,'' Alice said. ''A fancy name for 'hawklike.' ''What color are his eyes?''

''He's blind. He wears dark glasses all the time.''

''He's *blind*? They're never handicapped. He won't stay blind, you wait and see. That's just to get Nurse Prue to Ceylon.''

''He was her sister's fiancé but she didn't want him once he went blind. Nurse Prue went instead.''

''She couldn't bear to have him hurt.''

''That's it.''

Alice stood up. ''He'll get back his sight, don't you worry. Harlequin heroes are physically flawless — like Ken dolls.

Only in the newer ones they obviously have a cock. Which Ken does not. His shorts are welded on just so little girls will be safe."

"Why are they all so dark?"

"I don't know. It may be some kind of weird ethnic mythology. Look at how many are Spanish or Mediterranean. Latin lovers are more sexy. Something like that. They have blue eyes because there's a little northern blood in there somewhere. But maybe they also have black hair because it's such a nice contrast with the heroine's hair. The heroine usually has red hair, or red-gold. Sexy again. Lots of temper and fight. These girls are good girls with hidden fires." She smiled down at her daughter. "I don't know why I let you read such nonsense. Oh well, *Madame Bovary* is a good antidote." She paused. "Does your father know you read this crap?" She had trouble saying "Peter" but "your father" sounded awkward, as though he were a stranger, or dead. "Dad" sounded silly. Peter's parents, English, had referred to each other as "Mother" and "Father." Alice's parents, when they weren't fighting, called each other "Daddy" and "Muddie." Her older daughters had begun to call her "Alice" and if it still sounded strange, it was certainly better than never being able to refer to your mother or father by anything other than their generic or functional names. And yet she hated to hear small children call their parents by their first names ("we're all pals together"). The whole business of names was awkward. And who was she really? Flora's mother, Hannah's mother, Anne's mother, Peter's ex-wife, her wedding ring kept in a small square box at the back of her desk drawer. She had been going to give it back to him, to send it in a note which said "use it on the next one"—it had never been engraved—but somehow she couldn't do it. I must make a will, she thought, I must make it clear what to do with all the bits and pieces. What if she said, "I want to be buried with the ring in my closed fist"? Would everyone be too shocked or too saddened? She could write to Ann Landers. Had there ever been a letter on the etiquette of leftover wedding rings? There were probably hundreds of them right now, thousands even, just lying there in little velvet-lined boxes in the backs of drawers. ''Dear Ann Landers, I have a friend who has a problem — '' Signed ''Baffled in British Columbia.'' Ann would know what to do; she always did.

Alice moved into the shade and took out a small book called *A Spanish Voyage to Vancouver.* Since coming to the island she had been interested in the Spanish exploration of the northwest coast and she had stolen this book from the library years ago, declaring it lost and paying a fee. One day she would take it back, for they had probably been unable to find another. The island had a Spanish name, as did some of the others in this archipelago, and it amused her to think of these Spanish captains coming through Porlier Pass and finding many low, small islands which they ''had no interest in exploring.'' She tried to imagine them in their little boats, feeling the full force of the wind once they had left the shelter of the entry. All those men sailing the oceans of the world, claiming lands in the names of queens and kings, naming rivers, islands, mountains after saints and ships and fellow officers. Did their wives, when they had them (and if the men came home), hang on their tales of Indians and waterfalls, strange ceremonies and strange sights, the way Desdemona hung on the words of Othello? What if women had been the explorers? Would things have been different then? Imagine a ship of women then, trained in the use of all those wonderful instruments — quadrants, chronometers, the azimuth compass — knowing how to steer by the stars, knowing the difference between an artificial horizon and a real — would they have behaved any differently? Would they not have fired their guns and claimed things in the name of whatever monarch they served? Would they have offered beads and pieces of copper in return for sea-otter pelts? Would they have scorned young male slaves — excellent specimens — sent out to them in canoes? Probably not. It probably wouldn't have made any difference whatsoever. And yet.

But now a woman was getting ready to go to the moon.

It wasn't the same though — or was it? — shot like a parcel from the earth to the moon and back. Or like a bullet in one of those harmless guns you used to see for young children — a cork on the end of a string. Would Flora go to the moon — pop — like a cork on the end of a string?

The night of the first moon landing Peter and Alice and the two kids (not Flora, not yet) watched on a borrowed television. As the moment came closer for the actual landing they all went out on the porch, as though they would be able to see it — some change — a small pock — something. Nothing, of course; it was all too far away. Even with a telescope they would not have been able to see it.

On the blue screen the simulated landing module advanced inexorably toward its target. Alice felt frightened — it was as though the earth were doing something it had no right to do.

Nothing happened, of course. Except now there was a flag and a footprint. Up there, on the virgin moon.

Alice, restless and not sure whether she wanted to know, at just this minute, about a place so awful the Spaniards had named it Desolation Point, got up and walked to the very end of the rocks. Flora, she could see, was back on her stomach again but still reading *Nurse Prue*. It was probably a terrible thing not to stop her from reading such trash. The books were free at the store; there was a whole "library" of thrillers and romances left behind by summer visitors. All you had to do was return them. On the inside cover of some of them there was a garland in which you could write your name: "This Harlequin Romance is the property of ______." Who would want to admit to reading that stuff?

Still, interesting girls, those Harlequin girls. They often had careers at the beginning of the book but never at the end. And weak ankles. Always twisting their ankles as they ran through the woods, trying to escape their destiny. She stood at the point, looking out, seeing the noble Spaniards as they came along in their small boats, sailors rowing like crazy because the wind had suddenly hit them, the sails next to useless. But just at that moment what she really saw was a yellow canoe, with two people in it, coming this way, over to the Gulf side, heading for the bay. Alice squinted, took off her sunglasses then put them on again. It couldn't be. Although there was no reason why not. It was. It wasn't. It had to be.

The canoe was moving fast. Soon it would be right below her. She could see his red hair. Perhaps they weren't coming here. Perhaps they were just out for a ride.

Flora had come up behind her. "What are you looking at?" she said.

"The people in that yellow canoe." Flora was short-sighted, like her mother. She had taken her sunglasses off so that she wouldn't get raccoon circles around her eyes. She shook her head.

"I can't see a thing. What about them?"

The red hair—a burning bush. It had to be. But who was with him? It couldn't be Stella. This was 1979 and she was in Ireland, leaning against some crumbling castle wall, wrapped in a handspun shawl. "Come dance with me in Ireland." When Stella left him, Peter had rolled up all her dresses and left them by the front door. He had put her kittens in a basket and left them, with a note, at the university: "Here are your children." He had told Alice all this, sitting on her (rented) couch in the (rented) apartment where she was living for a year. To teach. To earn some money. To be near her children, her kittens.

"Do you remember Harold?" Alice said to Flora. "He changed his name to Gabriel at one point, but everybody was changing their names then. You're lucky, I suppose, you didn't get called Sunshine or Rainbow or Chanterelle."

"Chanterelle is nice," Flora said.

"Yes. An edible fungus.

"Anyway, do you remember that guy? He used to live with Stella. Before she . . . . When we all lived here together."

"I don't think so. I don't remember much about when we all lived here."

"You were very small." The yellow canoe had disappeared around the point.

"He's deaf," Alice said. "He's deaf and has a lot of red hair and used to live with Stella."

"Are you ready to go?" Flora said. Sometimes she reminded Alice of the cat. That same look of absorbed inattention. "Or can we have one more swim?" She remembers him, Alice thought, but she does not want to talk about Stella. Who hurt her father, her hero. She only wants to stay out of it, go swimming. Me too.

"We can both have one more swim." She pulled her T-shirt over her head. "Come on."

But as they ran down off the rocks and toward the sand the yellow canoe was just pulling into the lagoon. Alice hesitated. Finally, "You go," she said. "It is Harold. I want to say hello."

Flora looked again at the people in the canoe. "I'm going swimming," she said and ran off across the sand, away from the lagoon and toward the open water. Alice stood still and waited. ("Mum, is that guy stoned?" one of them had said. How easily they said it — "stoned." "No — or sometimes he may be. Isn't everybody these days? But that's not why he talks that way, he's deaf." 1972. "Mum, is that guy stoned?") If she ran away now. Why should the past rise up just now, just now when she was trying to be, was succeeding in being, so calm. "Calm as a clam," as a poet friend of hers once wrote. Clam down.

Harold stowed the paddles (no life jackets, Alice thought, of course not), looked up, saw her and waved.

So much for hawklike features. She stood there, waving and smiling, as the deaf man and his companion came across the beach.

Although they called it "the cabin" it was really a small, white clapboard house, or cottage, trimmed in red. The old man who had lived there and died there must have been extremely fond of red. The trim around the windows (and around each small square pane) was red; the crude benches on the porch were red; there was a strange old homemade lounger, mattress gone and metal springs covered in rust, set out under the apple trees; it too was painted red. After the new room had been added, the slope of its roof more or less the same as the slope on the other wing (which contained kitchen and utility room), it struck Alice how much the house looked like a broody white hen with a red comb. Later still, when they were given chickens, this just added to the image — for the chickens were Orpingtons, white, like the house, and with red combs. The house had no foundation; it sat low, directly over the rocky ground. The original structure was suffering from rot and there wasn't a straight line in the place. It was a little like living on a ship; parts of the floors tilted up, parts down.

They went to see other houses; they went to other islands. But that was a mere formality. "Let's take it," Peter said. "I'd love to fix it up." Which he did, made it livable, built a fireplace, found a pot belly stove in a junkyard in Dawson Creek. After three summers Alice said she wanted to live there all the time, forever and ever. "I'd like to say I'm thinking only of the children — how clear the air is, how simple the life, how good it will be for them to get out of the city — but I'm thinking of myself, as well. And of you."

"What would I do over there?" Peter said. "I don't think I'd want to commute every day."

"You could quit your job; you could paint and sculpt. We'd manage."

"On what?"

"We'll sell the house in town, put the money in a savings account or something. And I make a little from my writing. I may, as time goes on, make more."

Peter decided he wasn't ready to quit his job. He suggested (or did she? Later on Alice could never remember whose idea it was) that Alice and the girls go first, try it for a year, see how they liked it. Hannah would have to do correspondence courses until she was old enough for the high school on Saltspring Island. She might find that awful. Anne might hate the local school. Flora was so young she didn't care where she was, and there was a family with young children just down the path and across the road. But the other two? "You have this fantasy about the simple life," Peter said, "but complex people don't lead simple lives. You may be the one who finds it the hardest. Let's have a trial run. At the end of a year, if everybody's happy, we'll make it permanent." He would come over on Friday nights and go back Sundays. Meanwhile, they would build a new room, with a big bed and bookshelves and lots of windows. It could be their bedroom and a new sitting room combined. He drew a quick sketch, excited.

"Windows all along the side," Alice said, "or all along the side by the bed. Then we can look out at all that green. Front windows so we can sit up in bed and see the channel, watch the sun in the morning come across to us." Peter laughed. "Your wish is my command." He drew in windows, old ones with many small panes, just like the ones already in the house. He drew a bookcase to one side of the bed. He drew the pot belly stove — "we'll have to get a small cook stove from somewhere for the kitchen" — he drew Alice sitting up in bed and drinking coffee.

"Draw you beside me," she said. He looked at her for a minute, then smiled. "I'm already up," he said, "I've gone fishing."

Later Alice said, "It's so strange. I've felt so close to you these last few months. I saw the new room, the move over here, as something wonderful, a new beginning to our marriage. I guess what I saw as a sunrise you saw as a sunset."

"No," he said, "that's not really true. I didn't know what I saw. I just had to live one day after the other."

"Why didn't you *talk* to me? Why did you let us come over here and get settled? It's so cruel!"

"I didn't talk to you because I wasn't sure. I still wasn't sure."

"When I was in town at the end of July," Alice said, "I went to see Anne-Marie and I told her I was afraid you were in love with somebody else. She didn't say anything to me."

"I know, she told me."

"She *told* you! And you still kept quiet. You still let me keep on believing we were all doing this together! There's something really twisted about that. You and Anne-Marie having long talks about old Alice and what to do about her. Can't exactly put her down like an old dog but at least we can make sure she's over there on the island."

"It wasn't like that at all. We didn't know what to do, what to say."

"And all that wonderful erotic tension of a moral dilemma. I should imagine that Anne-Marie, at some point, offered to get out of your life forever. Weeping, she told you that you had a wonderful wife and three lovely daughters and she had no right to break up your home."

Peter got very red in the face and Alice knew she had scored a hit.

"Didn't she? Didn't she do the old self-sacrifice act?"

"What if she did? Perhaps she really meant it. You always see things in such a negative light."

"Listen, I know Anne-Marie as well, don't forget. I know how sorry she feels for herself. I know that all last spring when I was trying to write, she'd phone up and interrupt me and start crying, say how 'low' she was feeling, how she hadn't meant to phone but was feeling so low."

"And you told her, didn't you, to stop calling unless it was after twelve-thirty or she was about to commit suicide. Your bloody book was more important than your friend!"

"My 'bloody book.' Oh, that's interesting, that's really interesting."

"I'm sorry," he said, "I didn't mean that."

"You said it."

"We don't always mean what we say in anger."

Alice took a deep breath. It had not occurred to her that Peter might be jealous of her writing. He had always encouraged her. She felt as though she had been hit.

''Listen, Peter, Anne-Marie plays games. She has been going through a bad time but she also refuses to do anything to help herself. When I said that to her — about not calling unless she was about to commit suicide — I said it with a laugh, you know, but also to let her understand that she *could* call, if she were really desperate. I couldn't take the phone off the hook because the kids might have an accident or something—it has happened in the past—but I was, yes, desperately trying to have a few hours to myself while Flora was at day-care. What's *wrong* with that? Why should I feel guilty for wanting that? Why shouldn't someone who is genuinely a friend understand that?''

''When people are in trouble,'' Peter began.

''When people are in trouble of course they turn to their friends, but almost every day? At a time when they know their friend is trying to write?''

''You had so much — ''

''I'm glad to see you put that in the past tense. Had.''

''That's not what I meant. You still have so much. Your writing, your children, this place, this property. And I would very much like to be your friend. Maybe we can be real friends now, in a way we haven't been before.''

''I doubt it,'' Alice said. She swallowed hard. ''I was thinking the other day,'' she said, ''how different the world looks through rain than it does through tears. A landscape seen through rain keeps its shape—it's like a picture painted by the Pointillists only done with lines instead of spots. Quite beautiful. Through tears it's all smeared and blurred, it's — '' She stopped and looked down in amazement at her hands. They seemed to have a strange life all their own. They were wringing each other and wringing each other and wringing.

''I cry too,'' Peter said. ''New beginnings are always hard. Please, Alice, can't we be friends?''

''It wasn't Anne-Marie that turned you on,'' Alice said. ''It was dope.''

Peter shrugged, helpless against her darkness. Then he said, "Well Alice, why couldn't you?"

That first summer there had been very little money. Alice had baked bread and made pies and Peter had caught fish. The two oldest declared they were sick of salmon.

"Nonsense," Alice said. "Nobody gets sick of salmon. Nasty capitalist brats. Eat up."

"Could we *please* have a hot-dog roast tomorrow?"

"They have no taste," Alice said to Peter. "Wait until *all* they can have is hot dogs. Red dye number whatever it is."

But oh they were happy happy happy. Alice had made a blackberry and apple pie.

"Perhaps you'd rather have a Sara Lee frozen cheese-cake?" Alice suggested. They demolished the pie and grinned at one another like terminal heart cases with purple lips and teeth.

"I love it here," Alice said to Peter.

"I do too." He had bought a small rowboat called a Peterborough, clinker-built, and went out every morning early.

The next year Alice and the children painted it bright red outside then Alice drew a big eye on either side of the prow and wrote "ONYAME NYAA," which meant: "God looks after tailless animal." Alice collected African sayings and proverbs. Later she was to come across one which she almost sent to Anne-Marie: "Nobody teaches a Cat How to Steal," and one she pondered herself: "Tell Me, Who is to Blame? He who spread his mat across the Path Or he who Trod upon It?" "Peter," Alice had said, "Anne-Marie is really down in the dumps. Maybe it would do her good to talk to someone like you."

One night Alice had a dream of apples. She was holding Flora in her arms and people were pelting them both with fruit. She was trying to protect the child with her arms and at the same time avoid the apples which were falling on her own head. She was calling and calling for help. She wanted someone to bring her a yellow hard hat to protect her head from the blows. She ran up the steps of a house (awake, she realized they were the front steps of her grandfather's house) and weeping, pulled open the door with one hand. Inside was a long, carpeted corridor, like a hotel corridor, with many cream-colored doors leading off it. She could just see Peter and Anne-Marie disappearing through one of the doors at the end. She called out to them for help but they did not hear her.

When she woke up, weeping, she knew of course who the "other woman" was. And realized she had known it all along. Had seen it coming, like something far away down the road, coming closer and closer and closer. Peter and Anne-Marie. Of course.

Although it was early September, the night was warm and she had left one of the little windows open. Now she leaned out into the quiet moonlit night. What on earth was she going to do? Alice Agonistes, her children asleep in the other room, secure in the velvet dark while their mother leaned against the railing of the night and stared down into the great expanse of sadness which stretched before her like an endless sea.

An old man was moving slowly across the sand with a metal detector, looking for buried treasure.

They had come here so often. Spanish Banks. Where Captain George Vancouver met Captain Dionysio Alcala Galiano on Vancouver's thirty-fifth birthday. Peter and Alice were thirty-five. They were talking about making changes. Used to come with the kids. Two of them dipped in salt-water here for the very first time.

Christmas trees burning.

School fireworks displays.

Picnics and sand castles, marvelous Peter-directed sand castles, the envy of everyone on the beach. They told each other they could no longer imagine living away from the ocean.

Had they ever come here alone? Just her and Peter? She couldn't remember. It hadn't mattered. "Alone" was a word that they didn't—hadn't—used very much, a word that had dropped out of their common language, obsolete, no longer needed.

Today she had come right out and said it, because she was troubled. But why was it so hard for her to say, without feeling guilty, "I want to walk along the beach with you *alone*."

"So you never really deeply loved me," she said, two months later. Sitting on the front porch of the little cabin in the dark.

"No," he said gently. "But the potential was always there."

"The potential."

"For a real relationship."

"Then why isn't it there now! Why can't we try again?"

"No," he said, "It's too late now. It's gone."

Stella had lent Alice and the girls one of her books on lipreading. Now Alice began to recite the lesson she was on.

"fra - froo - fri - frou
fre - froi - frah - fro - fraw"

Alice said.

"'The shape of the sound of *r* after *f*, *b*, *p*, and *n*.'

"The bride wore a white veil.
Take a deep breath and hold it.
I will try not to break my promise.
A long branch of the tree broke off in the storm.
We must freeze — "

"*Stop* it," Peter said.

It was hard for Alice to see. Things kept getting in the way. The car was going either too fast or too slow. She tried talking about it in the third person. "Alice and Peter Hoyle were a happily married couple. A happy family just like the happy families in the card game except that there were three Misses Canada and no Master. But ever so happy. Everybody said so and Everybody ought to know. When Alice's friend Anne-Marie, correction Alice and *Peter's* friend Anne-Marie, came to their house to visit she usually called up after she got home. In tears. 'You are such a happy family,' she would say. 'I always leave your house feeling so comforted and yet so alone.' Anne-Marie was extremely beautiful, a real Grecian goddess and not like the Greeks you saw shopping in Kitsilano, with their tired black skirts and sweaters, their sacklike shapes. No, she was one of the nymphs on the Grecian urn, oh my yes. When she cried her eyes didn't even get red. She may even have wept pearls.

"Why is it so dark outside," Alice said. "The days are getting shorter. Soon it will be Halloween and we can all wear masks." She took up her narrative again. "Everybody felt sorry for Anne-Marie. Her husband was handsome, but mean. Nobody felt more sorry for Anne-Marie than Anne-Marie herself."

"I'm not leaving you for Anne-Marie," Peter said. "I nearly made that mist — I'm leaving you for myself."

"Did you know the word 'abandon' means to be set free?" Alice said. "Will Alice do that now? Call up her girl friends and sob into the dark telephone — boo hoo hoo? Will she begin by calling up her girl friend Anne-Marie? But Alice lives on an island now. There is such a thing as long distance. She will have to pay if she wants to talk to Peter. She could call collect, of course, he would have to accept the charges, he would have to accept the charges he would have — "

"Please stop!"

Alice pulled over to the side of the road, carefully so that they wouldn't go into the ditch. "I've stopped," she said, "now get out."

"My dear Alice," Peter said, "my dear friend."

"I'm not your dear Alice or your dear friend."

"I could drive if you're upset," he said.

"No," she said, "you must get out here. Right now. We started early; there will be lots of cars coming down. Hurry up—I'm in a hurry." She leaned across him and opened the door. "Get going."

He shrugged, reached into the back for his rucksack and got out.

"I'll call you tomorrow."

Alice slammed the door and made a reckless U turn, not looking to see if anything was coming. There was no radio in the car so she began to sing, as she had often done with the girls. "White Coral Bells," "Scotland's Burning," "The Ash Grove," "Maxwell's Silver Hammer." But singing by herself, singing to keep her spirits up, how small her voice sounded, so tiny and far away.

"I'll build a bamboo
Bungalow for two
Bungalow for two my Honey
Bungalow for two.
Walla Walla Walla"

She would have to pull off the road and stop somewhere or the kids would wonder why she was back so soon. Or Hannah would. The others didn't know yet. There had been no fights, no broken glass or accusations. It would come like lightning hitting a tree. The family tree. Ha. Ha.

Peter had wanted her to tell Hannah but Alice had refused. When the other two were asleep Alice had gone and got her and sat her down by the fire.

"Your father has something to say."

Peter had been sitting on the front porch, smoking up.

"The truth is," he said, "your mother and I can't relate to one another any more."

"Relate" was one of the new words, one of the words you absolutely had to use at least once a day.

Hannah just looked at them both.

"The truth is," Alice said, "your father has fallen in love with somebody else. The beautiful Anne-Marie."

Still Hannah sat.

"It's really not that simple," Peter said.

"It never is. And yet maybe it is. That simple."

"We're going to separate for a while," Peter said, "to see how things go."

"May I go to bed now?" Hannah said.

("You didn't have to tell me," Alice said earlier. "I knew. Wives always know. It has nothing to do with lipstick on collars or stray golden hairs. It's very subtle. The awful clairvoyance of the committed heart.")

But he wasn't leaving her for Anne-Marie, he was leaving her in order to find himself. He was willing to take the children.

She stopped at a small turnoff above the sea. They had come here their very first day on the island, looking for a house. Alice had made a picnic with cold chicken and wine. The real-estate lady was with them. She called them "kids." She used the word "cute" a lot and had a big booming voice. But they had found their house — or cottage rather. Deserted. A hole in the roof where a stovepipe had been. A smell of damp and animal droppings. The real-estate company had been planning to tear it down — to make the property more attractive.

Alice got out of the car and stared down over the cliff. The real-estate lady had called this place Lovers' Leap but when asked she didn't know of any lovers who had actually done it. Should Alice be the first? She got back in the car and drove home. She did not want to go back but couldn't think of anything else to do. She did not want to park the car, walk up the path, let the dog out, let the cat in, call softly, "Hi, it's Mum, I'm back."

But she did all that. Not turning on lights. Not going softly in to see if Hannah were still awake. She went straight to the front room, where the big bed was, put a piece of wood in the stove and got undressed. It would have been easier to send a policeman to the door.

"I'm terribly sorry madam, there's been an accident. Your marriage — "

Reaching backward for a chair.

"Dead? You don't mean dead?"

"As a doornail, madam—if you'll pardon the expression. It was very quick, you can be thankful for that. There couldn't have been much pain."

She got up again and moved on bare feet into the other room. Bolted the door.

Just in case.

Just in case he didn't get a ride and had to come back. Let him sleep outside in one of the sheds.

But he didn't come back. Stella and Harold, taking friends down to the ferry, saw him waving, stopped and picked him up.

Alice didn't know that, then, as she lay wide-eyed and unforgiving in the enormous bed, wishing he'd come back so she could tell him to go away.

SEPTEMBER

Just his luck to be picked up. But he's a lucky man, always wins at games, for example. I wanted him to have to walk along the dark road with his pack on his back — fol de ri/fol de rol—I wanted him to miss the ferry, I wanted him to call me from the pay phone at the south end, ''Look here, I've missed the ferry, do you think you could come down and get me?'' And I'd hang up, I wouldn't even bother with no, I'd just hang up. But quietly, so as not to wake the children.

Rage, rage, rage. So full of it. How dare he walk away! ''I'll call you during the week.'' I vowed I wouldn't talk to him but of course I did. And cried. ''Hello old friend,'' he said, in his soft, *caring* voice. But I'm not stupid, I could hear the ''end'' in ''friend,'' I could see the ''rust'' in ''trust.''

This was to be a commonplace book, where I would finally bring together all the words and definitions and phrases I have copied out for years on scraps of paper. But I need it now for something else, I need it to stay sane. It is much larger than the usual five-by-seven black-bound notebooks I work with but I need that bigness—I need to sprawl, to scrawl, to pull out from myself the great glistening sentences full of hate and fury and fling them, still wet and steaming onto these white pages.

A month ago, up on the ridge he asked me, Alice, what do you really want? I was so happy, so content, I answered ''Nothing.'' I'd read my Shakespeare — what did I think would happen if I gave him an answer like that? But I was feeling so good! I had on my violet sweater and a long red paisley skirt. I had on a necklace of rose hips and cloves. That will become now, in my memories, one more of my ''crisis outfits.'' I can always remember what I was wearing when a crisis occurred, a teal-blue velveteen party dress the day my grandmother had her fatal heart attack, a tartan skirt (Hunting Stewart) the night I lost my virginity. Why do we say ''lost'' when it is (usually) so freely given? A certain blood-soaked blue-and-white-striped cotton nightgown (why do we say ''lost'' when what is gone can never be found, is gone for good and you know it, you looked, you had the evidence of your own eyes). Every time I wear that outfit I shall hear Peter's voice, feel the warm sun on the back of my neck, smell the scent of cloves. ''Alice, what do you really want?''

Now I sit here in bed, wrapped in his Viyella dressing

gown, the cat at my feet, writing, writing, writing by candlelight so it won't disturb the children in the next room. Oh that I were covered in soft fur and could gather them under me, as Tabby does when she has kittens. Tabby never worries about "the father." And if I were a cat, like Tabitha, I could teach them useful, necessary things. I could teach them to pounce and tear, for example. I could teach them to pounce and tear.

"Alice," Peter said, "we've got to get things settled. About money, the children, visiting."

"Settled." For the first time she felt the full power of this word "estrange." Lowry and his "espiders." Alice and her estranged husband.

"Yes."

"'Everything ranged neatly on shelves,'" Alice said.

"What do you mean?"

"Something Doctor Aziz said in *A Passage to India*. About the English." Then, "You do it. You make the arrangements."

"No, we must do it together. Come on, just the essential things. I have to go in an hour."

On the way down to the ferry he said, "If I've hurt you, I'm sorry."

Peter painted nudes with loins of irresistible attraction, breasts like bloated wineskins, bursting to be touched and tasted, smelled and sucked. Waiting, voluptuous, lying on their backs and sides, with heads or faces hidden. Flesh so real it smoked. But passive, faces turned away, waiting to be penetrated. People always marveled at Peter's nudes.

"You could drown in the flesh of those women," a friend said once.

Peter gave him a drawing for his birthday. The friend

came the next day, laughing. "You know what my old landlord said when I asked him how he liked it?"

"What?"

"Just a twat looking for a hot prick!"

Peter and Alice laughed.

OCTOBER 10TH

(L. *volupt -as,* pleasure.—L. *volop,* adv. agreeably. —L. *vol - o,* I wish.)

When we got back to the cabin the girls were piling up the wood outside. I followed Peter into the bathroom and shut the door. We were both stoned. I said nothing, just took off my underpants. He unzipped his fly and we fucked, up against the bathroom door with the picture of those dear little Dionne quintuplets looking on. We heard the kids calling.

"I don't care. Shh. Don't answer, Peter. I love it. Oh harder harder. Oh my *God.*"

We stood exhausted, with our arms draped over one another's shoulders.

"I know it didn't *mean anything,*" I said, "but it certainly was nice."

We tidied up and went to greet the children.

"My legs are shaking," I said. He ruffled my hair.

What a marvelous pussy you are
you are
What a beautiful pussy you are.

"Has your name always been Raven?" Alice knew she wasn't supposed to ask a question like that. You also didn't ask people where they were from. If you did, the chances were you got an answer like this: I am coming from the Slocan Valley or Campbell River or the caves at Matala or some ashram in India, meaning that was the last place they'd been. (And might be movin' on again, move with the changes doncha know.)

But this bearded man with the soft brown eyes didn't seem to mind. He smiled and Alice saw that he had very few teeth. He was only about twenty-five and she wondered what had happened. The Indians pulled teeth out sometimes, rather than fill them, but it was rare to see a white man, especially so young and attractive and obviously in good health, smile such a toothless smile.

"My named used to be Wesley," he said. "That was cool, but I got tired of it."

The young woman with him, Selene, had all her teeth. Very straight and white — teeth that had been looked after since childhood. A soft voice. They both had soft voices. Soft voices were, of course, "in." Peter had met them down at the wharf as he was coming in from fishing. They were spending some time on Pylades Island and had canoed down to get a few supplies from the store. And just to enjoy the morning, maybe visit some friends at Coon Bay. Last year they had spent several months canoeing in Nootka Sound with another couple. Right now everybody was sitting on the porch drinking beer and smoking up. Hunh. Hunh. Hunh.

Hannah and Anne and Flora were down at the boat launch, playing.

Peter had caught a few grilse so eventually everybody had supper together. Selene and Raven were vegetarians — meat made you aggressive — but they did, occasionally, eat fish. Selene was very beautiful and had a certain calmness about her that Alice found enviable. She had come from New York City a long time ago but you couldn't tell that from her voice. "Soft, gentle and low," Alice thought, "an excellent thing in Woman."

They all shut their eyes and held hands before the meal.

''What wonderful people,'' Peter said, after they had left, promising to come back again. The girls had gone down to the wharf to watch them set off in their canoe. They had a small kerosene lantern to use as a running light if they needed it but the moon was full.

'''Sailed on a river of crystal light,''' Alice said. ''It's really that kind of night, isn't it? Shall we go for a walk or even a row?''

''I think I've done my rowing for today.''

''I could row,'' Alice said.

''Let's just sit. Let's not *do* anything.''

They were on the front porch. The moon had risen about an hour before. There was no wind. Later Alice said to Peter, ''Peter, do you love me?'' She knew it was another one of those questions you weren't supposed to ask.

He sighed. ''In your terms, I don't know. I just don't know anymore.'' He sounded so tired and worried that she put her arms around him and held him tight. It would all work out, wouldn't it? Weren't they lying here together? Weren't they man and wife? Weren't their three children peacefully asleep in the other room?

She had an image of Raven and Selene, paddling wordlessly under the moon, good people, people with soft voices who did not get angry or possessive or full of fear.

She did not know, then, that every time Selene got a letter from her mother she had an asthma attack. That sometimes they were so bad she nearly died.

She did not know, then, that Raven and Selene had their troubles too. She had not yet seen Raven, laughing, tip Selene's soup back in the pot saying Selene doesn't want to eat tonight. She had not heard Selene cry.

In her own self-centered distress she imagined that for people like Selene and Raven the world was paradise before the fall, that their nights held only bright, glittering stars and moons forever full, forever beaming. She imagined them pulling their canoe up onto the beach, walking hand in hand to their tepee, sleeping with their arms around each other, waking early and making love.

She did not know that both of them had already been married to someone else. That beneath the clear surface of their serenity were all kinds of things that wouldn't bear looking at.

She put her face against Peter's strong back. He was in love with somebody else but she didn't know what to do about it.

If she had talked to him, then?

If she had caressed him? Put her hand between his legs?

If she had said I really love you I just don't know how to get through to you? She, who was so concerned about Harold's deafness; she who was reading books on how to pronounce her words more clearly, kept silent, did not even try to speak what was in her heart.

Cordelia: Nothing, my lord.
Lear: Nothing!
Cordelia: Nothing.
Lear: Nothing will come of nothing;
speak again.

And if Peter had done that, asked her to speak again, would things have been any different? It had been two weeks since the beginning of term, since Alice and the girls had begun their new life. But not that new, really, since, so far, except for Anne going to school every day and Hannah beginning her correspondence courses, it seemed more like an extension of summer. "Nicer, in a way," Alice said to Peter, on the phone, "all the summer people have gone. I think the island, since Labor Day, has risen three feet out of the water." The weather held, and one evening Alice packed everybody in the car and went to Montague Harbor at the other end of the island for a picnic. The provincial campsite was practically deserted. The three girls went for a walk while Alice sat on the warm sand, her back against a log. In spite of her worries about Peter she felt very content. And very lucky. Everything would work out; how could it not? She wondered what Peter was doing, whether the house seemed very large and empty without them there. She wished he had decided to quit his job, but she wished it only in a vague way — because she knew he would enjoy this particular evening.

She would see him Friday nights, even Thursdays sometimes, if he could arrange it; and meanwhile she had the girls to see to, her book to write. Peter had said he wanted to have "intense relationships with other people," meaning women, meaning probably one woman, and that scared her. If she said, "If you do that, we're through" she had a feeling she knew what the answer would be. Their relationship wasn't intense. There was neither intense love nor intense hate. It was probably very old-fashioned. There was affection and respect. There was the love for their daughters. There was reading out loud to one another. There had been, all summer, the building of the new room. Their sex life had been more passionate lately, but different. This didn't necessarily reassure Alice. She remembered the wife of one of her professors in graduate school who had said she could always tell when he was having an affair with one of his students because, at the same time, he became very passionate toward her. And Peter had started going down on her—a new thing for him to do. She loved it; it excited her terribly. He laughed and said, "Now just go slowly, slowly," but she couldn't, with his tongue inside her lapping and licking. Thinking about it, with the water not quite still but gently lapping lapping made her want him right now. Wanted to say the magic words and conjure him up, to be there as the sun smoothly rolled down behind the hills. Wanted to be there on the warm sand, pull up her skirt, pull his head down, do the same to him, curve her fist around his cock, curl her tongue around it.

She saw the girls coming back, the two elder ones swinging Flora between them. And where would they be in her fantasy? Looking on? Giggling? Looking the other way? Never mind. Although it would be nice, right now, just to take off her underpants, walk into the water with her skirt held high, stand with her legs apart and feel those slow soft waves against her. She stood up and folded the blanket. ''Time to go home,'' she said. They drove up the island singing rounds.

Ten days later, Alice and Peter sat on a big rock just off one of the logging roads. They had left the children, Hannah in charge, and gone for a walk alone. Or not quite alone, for Byron, the dog, was with them. He had run ahead and circled back so many times that he had done the equivalent of three walks and was now lying down, tired out and panting, his tongue hanging out like a thin piece of ham. It was a beautiful day and they could hear the hum of bees in a meadow nearby. Alice sat, in a purple sweater and long red paisley Viyella skirt (''my hippie skirt'' she called it), content just to be in the sunshine, to be on a walk, to realize that the girls were happily playing without her. The night before, after the groceries were unpacked and the girls had gone to bed, she and Peter made love for a long time. He spread out his hand and passed it over and over her face, very gently, then fell asleep in her arms. In the morning, early, he got up and made the fire; but instead of going fishing, as he had said he wanted to do, he came back to bed. It was barely dawn and very still, just a few tentative queries from the birds.

They began to make love again, Alice on top. Flora, in her hand-me-down nightie, stumbled in just as Peter came. ''What are you guys doing?'' she said.

''Kissing and hugging,'' Alice said, and they made room for her in the bed.

''That feels good,'' he whispered to Alice, ''your hand between my legs.''

So that when he said, up on the ridge, sitting in the sunlight next to her, this man she was beginning to relax with, beginning to feel free with at long last, when he said,

''Alice, what is it you really want?'' she didn't understand the question. She was so full of contentment that she said, almost half asleep,

''Nothing.'' Because it did not seem to her at that moment that there was anything more to want. That to even contemplate wanting more was impossible, beyond her, greedy.

Even after all these years she is not sure what the right answer would have been. But even then she knew better than to give the answer she did. The sun, the lichen-covered rocks, the pale green flesh of the arbutus trees, the hum of the bees — everything conspired to put her off her guard. And Peter's come still inside her, his gentle fingers moving over her face.

Years later, over at Coon Bay with Flora, reading a book of West African proverbs, she came across one that made her sit straight up, as though she'd had a sudden pain:

''It is the path you do not fear that the wild beast catches you on.''

''You understand that I can't make love to you any more,'' Peter said, ''but would you hold me tonight?''

''All right.''

''Have you been making love to Anne-Marie for a long time?'' Alice whispered to Peter's back. She had her arms around him, tight.

''It depends what you mean. If you mean just screwing and fucking, no. But it's more than that, it's so far beyond that.''

''Why no fucking?''

He was silent. Then, ''Let's get some sleep.''

''Why no fucking?''

''I can't.'' He began to cry.

''Oh Peter, did I do that to you?''

''I don't know.''

Is that why he had to leave her? To find out?

After he was deep asleep Alice lay there, very gently running her fingers up and down and around his body, memorizing him.

Wide-eyed, wishing the morning would never come.

OCTOBER

Adore (L.) L. ad -orace, to pray to. -L. *ad,* to; *orare,* to pray, from as (gen. or - is, the mouth). cf. *Oral*

Where the bee sucks there suck I. "This was to be a commonplace book," but what could be more commonplace, these days, than the breakup of a marriage? "I *hate* marriage," Peter said the other day when I was in town. "The institution that is." I had a sudden vision of other institutions I have known, of barred windows, of somebody banging on a door with bloody hands. He had taken me to lunch at the Faculty Club and we were pretending to be friends. I asked him how he was going to introduce me from now on, since my name was the same as his. "I suppose you could just say, 'this is Alice,' a lot of people only have one name these days. Or I could change my name to Alice Apple or Alice Blackberry, something like that. Or you could change *your* name. When people are born again they often do. Saulus/ Paulus. The hippies didn't invent that game. I wonder what would be a good name for you? Besides Traitor," I said, "besides that obvious one."

He doesn't get angry; he just smiles his forgiving smile. I am "laying another trip" on him but that's cool, he can deal with it. We go to Woodward's Food Floor and buy groceries for the next week. We have decided he won't give me a regular allowance — I want to try and see how I can manage on my own. If he contributes groceries. If he pays the medical insurance, pays the dentist. We don't want to bicker about money. After all, as he says, that's not the kind of marriage we had. I murmur no, no of course not, grateful, *grateful*! that he can say something positive about our marriage. Now I sit here in bed, radio turned low, candle flickering, wondering why I don't get a good lawyer and hit him for everything he's got. "We didn't have that kind of marriage. . . ." What kind did we have, then, tell me, you asshole. Here I am on this island with three kids and a drafty house with no inside doors, no *privacy* for any of us in our shock and grief. Only Flora is unaffected, or seems to be. The rest of us avoid one anothers' eyes. Anne goes to school on the bus every morning early and has already made friends; she'll be okay, at least for the time being. Flora goes down the path to the baby-sitter for part of the day and doesn't seem to be aware, so far, that anything has gone wrong. But Hannah and I sit, each at one end of the kitchen table, heads bent over our work, unable or unwilling to talk about it. I hate him. I hate the cowardice of it all. I have been reading Captain Vancouver's journal. Oh for some of that eighteenth-century discipline. How I should love to see some flogging done.

"May 23. Punished James Bulton and George Raybold the first with 36 and the other with 24 lashes for theft." Somebody else gets twenty-four lashes for insolence. Somebody else gets twelve for "Disobedience of Orders"! A man named John Thomas gets thirty-six lashes for "Neglect of Duty." Sometimes he refers to a flogging as a "slight manual correction." Oh to give Peter a slight manual correction, or even a severe one. Bringing the lash down on his shoulders again and again until he begged for mercy. And the beautiful Anne-Marie. Would I have her flogged as well?

And yet in real life, or unreal life, whatever you want to call this thing that is going on, I got stoned with him this afternoon, watched a bumblebee and asked him to make love. And now, I suppose, he is sitting with Anne-Marie, or lying with his arms around her, telling her all about it, how it really didn't mean anything.

If I were going to change my name I'd change it to Midnight, to Hecate, to Amanita, yes to Amanita, beautiful and deadly.

Take and eat this. In remembrance.

Peter said that perhaps you are only "in love" once, in that absolutely bowled-over way. He and Alice had been smoking up—or he had, and Alice had taken a toke or two.

Then they went outside and leaned against the fence. It was very warm, summerlike, and yet it was the first week in October. Next week they would pick apples. She had thought of hiring someone this year but he said no, he'd come over for at least a day and a night. A fat bumblebee lit on Alice's bare arm. "Isn't it beautiful," she said. She stared at it, the fat fuzzy body, like some miniature child's toy, while Peter began talking about men in prisons using inflatable dolls for sex. The conversation seemed silly to her — what did she care about men in prisons or what they did for sex? It was far more interesting to watch the bumblebee, who did not seem to notice that he was sunning himself on a human arm. "Their real name is humble-bee," she said.

But then she too began to buzz; it was as though the bee had injected her entire body with a soft, insistent buzzing. She gently shooed the bee away, saying, "Peter I want to make love."

They walked around to the back road and over to the next property. Full of tall grasses and not yet built upon. Peter did not look at her but watched himself going in and out, in and out, gave a sleepy smile from time to time. "The sun feels so good on my ass," he said ("my awhs"). It went on and on, or seemed to, her body humming more and more in just one spot, the bee at the very center of the rose, there and there and there and there don't stop.

Petals folding in again, softly, voluptuously.

"Did you like that ?" he said. Her body ticking, ticking like the metal in an overheated car.

Then later told how he and Anne-Marie had been talking a lot about love, how she had copied out some ee cummings sonnets which turned out to say it all so much better than either of them could. Alice had to understand that although what had happened back there was nice, he and Anne-Marie had something higher.

The little serpent who was always coiled up at the back of Alice's tongue darted out:

"You mean newer," it said. Peter gave her a forgiving look.

"No, I mean higher."

"'Higher,'" the serpent said, "what do you do, stick it in her eye?"

He had told her he was impotent with Anne-Marie. Somehow Alice felt that was all her fault. Now she imagined them clasped in each other's arms, spirituality flowing between them, and desire, but unable to consummate that desire. What did that feel like? Did he go home and masturbate? Did she, after he had left, lie on her nice couch by the side of her fireplace, thinking of Peter and rubbing herself with her fingers?

She never found out (didn't ask, didn't want to know) if Anne-Marie and Peter ever managed to fuck. They never did live together. Peter found another friend and then, eventually, Stella.

But before that, back in the early days of their separation, they sometimes went to bed together, overwhelmed by desire, overwhelmed in a way they hadn't been since the early days of their relationship. And sometimes not even "to bed" or lying down in a meadow. Once, with all three children in the new room, following each other like two cats, into the bathroom quick now shut the door, Peter's pants around his ankles, up against the wall, fucking, Alice crying out in pleasure against his shoulder.

She was not just a substitute for Anne-Marie; she knew that absolutely. They never talked about it and after a while it didn't happen any more. Or didn't happen to Peter. Logs that fall together and flare up, just as the fire is going out—was that it? Peter did try to make sure she understood that it didn't *mean* anything. "That was nice," he said in his new voice. (But part of her never believed him; a part of her felt that he was saying "I love you" and asking for help. That may have been wishful thinking; she would never know.)

When Alice came back from her walk Peter was mowing the lawn. They went inside together and Alice started to cry again. She sat at the big table and cried and cried.

Anne came home early. "There was no lesson. I went all that way and missed school for *nothing*." She slammed her clarinet case down on the table. Then she really saw her mother.

"Are you sick? You look terrible. Your eyes are all swollen up."

"I hope it isn't anything catching." Alice went into the bedroom and put a pillow over her face.

Peter knocked gently on the door and then came in.

"I'll take her out to dinner on the way to the ferry. It's better that way. We'll see you tomorrow." He kissed the top of her head. After she was sure the car had driven away she turned on the radio very loud and howled like a forsaken dog.

OCTOBER

I was coming back from a walk when I saw Selene and Raven's canoe pulling in to the government wharf. I panicked and went back the way I had come, sat on the steps of the community hall for a long time and told myself not to be silly. I didn't want to have to tell them, in front of the girls, that Peter and I had broken up. (In case I "broke up," in case I started to bawl.) And I felt that they might tell me I had to "move with the changes" or some such reassuring hippie aphorism.

To tell them that Peter has left me seemed to me to admit that I had failed. And so I sat on the steps of the hall until I was thoroughly chilled, until I figured they'd bought whatever it was they wanted from the store, had a visit with the girls and shoved off.

I came home along the back access road so I didn't know they were still there. Raven was chopping wood and Anne was stacking it; I could see Selene and Hannah through the kitchen window, laughing. Hannah laughs so seldom that it shocked me. It turns out they are spending the night and then going into Vancouver for a few days to have a break. They have called Peter (I'll bet they didn't call collect) and he said he was delighted to put them up. They didn't ask me anything so either the girls told them or Peter did, on the phone. We had baked beans for supper and Hannah and Selene made corn bread. Afterward we sat on the carpet in the front room and I read everybody "The King of the Golden River." Now everyone is asleep, or at least in bed, Selene and Raven in the shed. (They find the front room too hot, with the stove on, even though I let it go out when I go to bed.)

They make me uneasy, especially Raven. With Selene it may be a case of simple jealousy—good people, really good people, are not only hard to find, they're hard to take. I honestly don't believe that Selene has the vipers and toads I have inside my soul. But Raven says weird things. I don't know whether they are intended to shock or whether he really believes them, but they all relate to his belief in karma or the working out of one's fate. He says, of the Manson murders, that it was the victims' karma to die and to die in that way. It isn't so much that he condones it as that he "understands" it. I feel there is enormous anger in him, somewhere, and the karma business partly is a way of justifying his angry feelings. They aren't really his, they belong to a past over which he has no control.

And all these soft voices! Almost like the soft voices of psychopaths, inflectionless, menacing: "Now don't turn around and don't look back and everything will be all right." (And if you don't obey I'll blow your head off.)

Alice often slept in an old plaid flannel shirt of Peter's, one that he had thrown in the ragbag because it was beyond mending. She had taken it out and washed it and wore it when she was feeling low. One weekend when the children were over, Flora, restless, slept with her. Alice had forgotten about the shirt, which was under her pillow. When she plumped the pillows up and crawled back into bed with her morning coffee the little girl said, ''Why have you got daddy's shirt under your pillow?'' Alice made some excuse about being cold one night and grabbing it from the ragbag. The child seemed satisfied but Alice hoped she wouldn't tell her father. He would be exasperated, not flattered. He wanted her to forget him, to ''put him out of her mind.''

''Adieu, Adios Amigo, Arrivederci, Ciao.''

''I love you,'' she said, and kissed the lapel of his dressing gown.

''What are you crying for,'' he had said. ''Fourteen years as vegetables! My God.''

Once when she cried out, ''You are trying to destroy me!'' he replied—''Oh no, you were destroyed before I ever met you.'' Was this true? She thought about it. Chipped and cracked, yes — *crazed*, if you like, but not destroyed, definitely not beyond mending. She had seen herself once, when thinking on this, as a well-beloved tea or coffee cup—handle glued on with china cement, long scars in the glaze, maybe a chip or two on the rim, but serviceable. Perhaps to be handled with a certain amount of care? But that was part of it, he said so. ''Am I supposed to wait around while you go down inside and confront your demons? I have a life *too*.'' They had just been through a terrible afternoon together, shortly after they had parted. She was coming in on Fridays to buy groceries and he would meet her at the ferry, take her to the shopping center where he paid for the stuff (after all, there were three children to consider) and then they would talk. She wanted to understand how all this had happened, how it had transpired that he felt so strongly the marriage was gangrenous, incurable, that he could amputate it in one day, one conversation in the back booth at Lindy's Restaurant and Delicatessen (''The Corned Beef King of Vancouver'') over a cup, then two, then three, then four, of coffee.

When she came into the terminal he was talking to a

young man from Quebec. "D'you happen to have a two-dollar bill?" he said. "This fellow wants to get over to Saturna." Of course she did, and of course she handed it over, but she resented being put in that position. If she had said no, Peter, the new Peter, would be confirmed in his belief that she was ungenerous, "uptight," still "locked into the Protestant work ethic," or whatever it was the freaks said. But she resented handing over the two dollars. Aside from the groceries every two weeks, she did not ask Peter for money—she did not want it to turn into any kind of blackmail. She saw the face of the young man, smug, traveling light with his ancient rucksack and cotton sleeping bag, going off to "crash" with friends on Saturna. Who needed money these days when the younger intellectuals like Peter carried so much guilt about having a steady job — "*Merc là,* man. Merc là," he said, and hurried off toward the ferry. "Thanks man." (To Peter, who had been his go-between, not to her. Peter of course did not happen to have a two.)

"Far out," Alice said under her breath. "Vive le Québec Pauvre. The counterculture has bridged the language barrier."

"What?"

"Nothing."

Then as they were going through the terminal doors they saw a couple who were building a cabin at the north end of the island. They were about fifty, both very fit, very active. They had bicycled across Canada in their forties, camped on the beach in Mexico, raised two children and very obviously loved one another. He was carrying a mattock and she pulled the ubiquitous wheeled cart loaded down with groceries and with a spade tied to the top. They were going to retire and live on the island full-time. Alice remembered all the trips she and Peter and the children had made, leaving the car in the parking lot, staggering under supplies, garden tools, cans of paint, bedding plants, roofing. Peter unlocked the door on her side and when she sat down she started to cry.

"What are you upset about?" he asked, giving her his new distanced look. She remembered the psychiatrist she had gone to years before, his unctuous voice. "You are quite upset?"

"Oh." She sniffed loudly. "The Bedworths. Going out to work on their cabin together. So happy."

"But *look* at them!" he exclaimed. "Look who they are!"

"Who are they?"

But he had started the car and was lining up for the tollbooth. She knew what he meant, anyway. They weren't glamorous, passionate, romantic. They were good companions, that was obvious, but practical people who had married for life and perhaps had to keep busy on projects in order not to realize what they'd missed. That's what he meant, of course. She hated him for his easy condemnation of people he hardly knew. And yet she did that sort of thing—she had just done it in her head to the boy from Quebec. She was horribly depressed.

They did the shopping and he asked where she'd like to go for lunch. They usually went to the Faculty Club because he could sign for it and they had good soup on Fridays but she didn't feel like going there, she didn't feel like going anywhere.

"Let's go to the house and talk. Maybe order a pizza."

"All right."

They sipped beer politely until the pizza man had come and gone. Peter produced a salad as well. The house seemed terribly empty — she guessed he didn't spend much time there any more. He rolled a joint.

"Give me a puff," she said. Then, "How's Anne-Marie?"

He held the smoke in and sniffed — hnh, hnh, hnh — (Anne imitated this beautifully, daddy smoking up), then, in his smoky voice —

"Do you really want to know?"

"I guess not. Is she planning to move in?" Her voice was a bit faraway and certain warning bells rang in her mind, certain faraway signals flashed, "get off the track."

"No, she hasn't moved in, she isn't going to move in." (Hnh. Hnh. Hnh.) (Quietly) "She cares about you, you know."

Hnh Hnh Hnh. Alice, in *her* smoky voice, in spite of the louder alarms, the brighter lights. "Anne-Marie is sick. I wouldn't be surprised if she loses interest in you now that it's not a big romantic secret any more (Hnh Hnh Hnh) Anne-Marie only wants what she can't have. D'you know Saint Augustine's *Confessions*?"

(Hnh Hnh Hnh) "No. No, I don't." (Voice very soft now. New hippie voice.)

"He talks about a pear tree he and his friends used to steal from. Just for the theft. The pears were bitter and they didn't want them anyway.

"I'm working on a little story about Anne-Marie," she lied. "Anne-Marie and her penchant for husbands. I'm calling it 'The Pear Tree.'"

"You always see the worst in people," Peter said. He stubbed out the joint, put it in his pocket. The pizza, with three slices gone, looked like an illustration for fractions from her grade three arithmetic book.

"I was never any good at that," she said, staring at the pie.

"What?" He was away, with a silly smile on his face.

"Fractions." They both stared at the pie and began to laugh.

"Algebra was worse — ladders leaning against walls at absurd angles."

"Tanks filling up and emptying."

"Trains headed straight for each other at different speeds." They were both laughing, giggling. She put her hand out, touched his knee. He picked it up gently, looked at it, then placed it back in her lap.

"Don't start."

She looked at him and he had, magically, moved back twenty or thirty feet. There was an enormous distance growing between them. Like an earth fault, suddenly opening up, he on one side and she on the other. All kinds of things falling into it. Alarm bells and flashing lights. Warning shouts. Scientist being interviewed on the national news. Saying, "I told you so."

''No!'' Then, in panic, ''I shouldn't have smoked that joint!''

He moved another twenty feet away, his voice barely audible — ''Just calm down, don't be silly.''

She put her hands out to him, across the horrible canyon that divided them. ''I shouldn't have smoked that joint!''

Then her arms, stretched thin as spaghetti, thinner than wire, touched him, her hands disappeared. She was horribly frightened.

''I can't get back,'' she said, ''I can't get back.'' She had become unreal. She could pass through people like a ghost. It should have been fun — she had always envied Invisible Scarlet O'Neill. But it wasn't fun at all, it was horrible.

Terrified, she got up and began to walk up and down on the braided oval rug, making certain patterns which her feet seemed to know automatically. Figure eights, pentagrams, strange forwards and reverses.

''I'm over the edge, Peter! I'm over the edge!'' He sat there, a tiny figure, so far away. A little mannequin.

''I don't know whether the best thing I can do isn't to leave you alone,'' the voice said.

''No! Don't leave me!''

''I'm going in the kitchen to make some coffee.''

''Don't leave me.'' Up, down, around that bit, then back, start again, don't stop. She daren't look at the table. There were tiny snakes of a bright-green color — green mambas no doubt, just hatched, crawling all over the pizza. Did they have venom when they were that small? Better not to take chances. Were they *neurotoxic* or *hemotoxic*? Could she remember what to do? All that stuff she read up on in her father's zoology books.

''Don't leave me!'' Five six seven.

''You know where the kitchen is.''

She rushed after him. ''There are snakes in the other room!''

''Only in your head. Try and think of something good, switch channels.''

''What? What good?'' She stood in the kitchen wringing her hands. Her fingers reminded her of snakes. And yet her hands seemed so far away. If her body went over to Peter's side, where would she be? Perhaps if she screamed. She closed her eyes and began to scream and scream. No. Someone else was screaming for her.

"I'm not crazy," she said, "it's the dope."

"I know you're not crazy, we both know it." Suddenly he sat down at the kitchen table and began to cry.

"What? What are you doing?" He was his right size again, sitting with his hands over his eyes, weeping. He who never cried.

She, sobered, went to him and put her arms around him, her real arms. He wept into her strong arms.

"I don't know what to do about you," he said.

She shuddered and shook her head to clear it.

"It's not an act," she said.

"I know that, but I don't think I can help you any more."

"I don't think you ever could," she said. They drank cup after cup of coffee and then she lay down for a while and then he drove her to the ferry.

"I'm all right now," she said.

"I know you are." He could have added what she knew he was thinking, "Otherwise I wouldn't leave you alone with my precious children."

Months later, when she was in town by herself, tidying up the sitting room, she found two pieces of paper. One, which must have related to the weekend he punched her in the face, was the beginning of a letter to a friend:

"I blew it. She blew it. I guess we all blew it. I want to write and tell you about it, why I hit her, but took psilocybin a little while ago—I think I went back to the womb and was kicked out again. Now I'm watching the pen make all these lovely letters. Watching my hand move — what miracles! The crossing of a *t* the dotting" — that was all. The dotting of an eye.

The second was ripped from a notebook:

"If I help her destroy her demons, will I destroy her

angels too?'' Rilke, she thought. The first present he had ever brought her was Rilke's *Letters to a Young Poet*. He had hidden it under her pillow before he went off to catch the bus to his parents' home.

She looked at both pieces of paper and then folded them up and stuck them under a pot on the mantelpiece. He had moved out and left them behind for anyone to see. They had no meaning for him any more.

Peter phoned. He wanted to know the children's birthdays for the medical plan. He was laughing, sheepish.

''I do know the months,'' he offered.

Alice held the telephone away from her ear for a moment, stared at it thoughtfully.

''Hello, are you still there?'' How English he still sounded, after all these years.

''Find out for yourself,'' she said, and hung up.

Later I thought about the word birthday. Birth. Day. The child is told that it's his birthday, but it's the mother who knows. And what she was wearing. And what the weather was like. And the incredible sense of relief as she lies there, bleeding, perhaps torn, smiling. And the right number of fingers and toes, has he? Nothing missing? Nothing too much? Truly the birthday belongs to the mother, not the child. Happy birthday to you. The child blows out the candle flames; it's a game. The mother remembers how it was, how it seemed as though the child might be scalded by the blood which boiled and bubbled between her legs.

If the child is a girl they count the number of candles still alight. That is the number of children she will have.

Afterward there are games. Musical chairs (someone must always be left out, that's the point of the thing), the farmer in the dell (the cheese stands alone, the cheese stands alone), Blindman's buff. THANK YOU VERY MUCH I HAD A LOVELY TIME.

> Then may be said: Grant, we beseech thee, heavenly Father, that the child of this thy servant may daily increase in wisdom and stature, and grow in thy love and service, until he come to my eternal joy; through Jesus Christ our Lord. Amen.

> The Woman, that cometh to give her Thanks, must offer accustomed offerings, which shall be applied by the Minister of the Church-wardens to the relief of distressed women in child-bed; and if there be a Communion, it is convenient that she receive the Holy Communion.

That's called the Churching of Women. An April day in England. The young priest moving awkwardly up and down the rows of women, sensing our communion to be so much stronger than what he held out in his silver bowl. "The blood of Our Lord Jesus Christ which was shed for thee—" My mind turning more to Mary. After the Immaculate Conception, the maculate delivery, pushing and grunting and lowing like the rest of the beasts in the manger. But the young priest walked among us talking about the blood of Jesus, asking us to take a sip, don't be greedy now, wiping the lip of the chalice with a linen napkin, on to the next. Could he not smell the blood scent in that room, the women who, under the crisp white coverlets, were placidly bleeding?

And when he delivereth the Bread, he shall say,

The body of our Lord Jesus Christ, which was given for thee, preserve thy body and soul unto everlasting life. Take and eat this in remembrance that Christ died for thee, and feed on him in thy heart by faith, with thanksgiving.

And the Minister who delivereth the
Cup shall say,

The blood of our Lord Jesus Christ which was shed for

thee, preserve thy body and soul unto everlasting life. Drink this in remembrance that Christ's Blood was shed for thee, and

be thankful.

(said to the
partakers of his most blessed body and blood.) Why did we not put our hands between our legs, show him our blood. "Drink this in remembrance."

For the word is not made flesh, it's the other way around. The flesh made word, or rather, in the begining, made cry, made howl. *Then*, later, the word, simple at first, all babies make it, race, color, creed do not come into it. Made word: Ma — Ma, Ma — Ma, the breast. And the breast, like the magic pitcher in a fairy tale, empties and fills up again.

Take this
Take this
Take this
In remembrance of me.
(and be thankful)

Alice and Peter met Stella through Stella's father, who had a beautiful piece of property just two miles down the road. He stopped them one summer morning in the store by the government wharf. He always swayed a bit, as though his body were subject to some wind that others did not feel. Later they were to find out what wind that was.

"Come for dinner tonight and meet my daughter. She's here from California. We'll have a salmon."

"All of us?" Alice said.

"Why not?" That was one of his favorite expressions. Another was "Are we in business?" A tall, thin, white-haired man, swaying slightly, smiling.

And so they all went, except the dog Byron, who howled behind the windows as they drove away. It was no good taking him; he was not one of those dogs at whom you snap your fingers and they lie at your feet the rest of the visit.

"Byron's sad," Flora said.

"Uh-huh. We'd be sad if we took him; he'll calm down."

They could have walked but they drove, because Peter had been digging a drainage ditch all day and was tired. If they walked, on the way home he'd have to carry sleep-heavy Flora on his shoulders. Alice sat in front with Peter, the three girls behind. A happy family.

Stella was small and olive-skinned. She had dark hair but it had been hennaed and was now growing out. Stella's hair reminded Alice of those hens with black and rusty feathers. Stella had her father's brilliant smile.

"Excuse my hair," she said from the couch. "It was one of my mistakes. My brother Glenn says I look like an Italian whore." It was obvious this image pleased her. Alice looked up at the enormous hand-tinted photographs in oval wood frames on each side of the piano. Great-grandparents probably. She fancied the great-grandma drew herself up a little straighter in her high-buttoned shirtwaist on hearing the word "whore."

It was a pleasant meal. They sat on the deck which ran along the side of the house overlooking the channel, ate salmon and drank a lot of wine. Alice noticed that Stella did not help with any of the cooking or serving but it was clear that she and her father were very fond of one another. About halfway through the evening a tall, very beautiful young man appeared. Glenn. He had a smooth, almost sexless face, "like an angel with grannie glasses," Alice said to Peter later. He spoke very softly, as did Stella, only Stella's speech was rapid. Her sentences were more like urgent messages over a bad connection on a long-distance call.

When Stella leaned backward to flick her ash in the ashtray her belly button showed and her smooth, just slightly rounded belly.

Stella and Glenn began talking about a lake they had discovered. There were three of them, quite close together, small blue oval marks on the island map but they'd never been able to find them. Then one of the Indians on the reserve had told them how to get there and they'd gone that afternoon. Only one lake was left, and it wasn't really a lake, it was a bog. "Like dancing on a sprung dance floor," Stella said. You had to walk in off the Coon Bay road, along a narrow path. But the water was lovely. There was sulphur in the water and it made your skin look wonderful. And you could swim there nude. Virtually nobody knew where it was. They would all have to go together, maybe tomorrow. Alice thought of Stella's smooth, slightly rounded belly and then of her own, which had seen better days, ribbed and streaked with silvery stretch marks like a white beach in the moonlight.

"Will you be staying here long?" Alice said. Stella shrugged. "I'm not sure. Dad says I can build a dome on his property if I want to. We'll see."

She had been in California for some years. She had been part of the whole Berkeley scene. She didn't say it but she didn't need to.

The next day Glenn and Stella came for a visit and asked if anybody wanted to go swimming. Alice and Peter had been working on the new room while the older girls were down at the boat launch, swimming and keeping an eye on Flora. Alice had been standing on a chair for what seemed like hours, holding up pink insulation while Peter stapled it in place. Her arms hurt.

"Why don't you go?" she said to Peter. "Take Hannah and Anne, if they want to go, and see if it's okay for Flora." She and Peter always made love in the dark. Her choice. Or was it? While she was making supper she gave herself a good talking to but it didn't seem to help. Stella was nice. She was naturally sexy in a way that she, Alice, was not. So what? There was something in Stella's eyes that looked like pain. She had seen things or done things that had hurt.

"I am very jumpy just now," she said to the cat, who was waiting impatiently for a cod's head to cool down (Flora was playing with Lego in the other room), "because I feel Peter slipping away, I feel it. I think I know what it is but I don't want to say it out loud, not even to you. And I don't know what to do about it.''

Alice put the cod's head in a dish and took it out to the front porch, the cat snarling with pleasure. Then she peeked at Flora and went back to the kitchen to pick all the bones out of the fish. Peter had gone out in the rowboat at 5:00 A.M. and come back triumphant. A ling cod, the very best eating. Her Provider. Her Good Provider. She had made bread and an apple pie from the transparent apples. They were a good team. Good companions. A companion was someone you (literally) broke bread with. She was content — most of the time — but lately he was not. She went outside to get some parsley for the stew she was making. The wasps were bothering the cat. Flora had been stung on the eyelid a week ago. It was hard to think of wasps as God's creatures; they seemed to be naturally vicious. When she was a child, at her grandfather's summer place, they had gone around every summer sticking lighted cigars into the openings of the gray papery wasps' nests which hung from the eaves. She and her sister were allowed one quick puff on the cigar first. She would have to try and find this nest and get rid of it. Was it true that only the females could sting?

''I don't know what I can do for you just now,'' she said to the cat. ''If I bring your dish inside they'll just come in.''

(In England Peter's mother would catch a stray wasp in her hands. ''Snip, snap,'' she'd say. ''You have to be quick.'')

Peter and the girls came back after Alice had fed Flora and just as she was thinking of eating. The girls had taken their bathing suits — had he? The rice was getting cold.

She heard the car door slam and the shouted good-byes and thank-yous.

''Oh Alice,'' he said. She could tell he'd been smoking up. ''You'll have to see it for yourself; it's magical. Great yellow water lilies, dragonflies, the cliffs up above.'' He laughed. ''It was hard to come away.''

Alice began to serve out the stew. ''Did you like it?'' she said to her daughters.

''I didn't like the walk in,'' Anne said. ''Lots of black ooze and roots to stumble over. Branches. But once you're there it's neat. Scary though. Dad and Glenn are going to build a platform we can sit on.''

''And you?'' she said to Hannah. Had they all swum nude? Peter was a very private person. Had she seen her father's body for the first time? They were all very private people — ''prudes'' was probably the word.

''It's very beautiful.'' She looked at her mother. ''Will you come with us next time?''

''It will be nice for you if Stella stays,'' Peter said. ''She's interesting. You'd have a friend.''

''We've got to do something about those wasps,'' Alice said. ''They'll drive me crazy.''

And the next time they saw Stella — or the next time Alice saw her — she had acquired a boyfriend. Alice and Peter were sitting on their new gold carpet, admiring the new room.

''This is Harold,'' Stella said to Alice in her quick, small voice. ''He's deaf.''

She and Harold had eaten some mushrooms earlier that day and then gone sailing. Harold, using his arms freely, described how the waves came up.

''Oh. Really wonderful. Such excitement.''

''Weren't you scared?'' Alice asked Stella.

''Terrified. But Harold loved it.''

''Love it rough. Wonderful. Lean backward. Earth tilts. Forward. Really Far Out.''

''He's what's 'wonderful,''' Alice said afterward, imitating his voice. ''All that red hair. And that voice. The Frog Prince.''

''They're both quite incredible. Have you seen the shack they're going to live in?''

''Not yet. Have you?''

''I walked up there the other day. It's right on the little bay. A beautiful location. But falling to bits. Harold intends to do a lot of fixing up and she's already scouting the Sally

Ann. They've been getting some nice old weathered boards from a cabin that's fallen down somewhere. They plan to redo the inside."

"Do they have to check with the Indians first?"

"I expect the band is glad to have somebody fix the place up. There's only a dirt floor in the front room; they'll have to cover that."

"I wonder what Stella does for money," Alice said. "Or what she will do now that she's staying."

"I think her father's helping her out for the time being. Her brother's staying too; he wants to build a houseboat. I might help him on the weekends."

"Haven't you had enough building for a while?"

He laughed. "Well, we'll see."

"And all the time he was running into town for 'materials' he was also having long tearful discussions with Anne-Marie. 'What shall we do about Alice?'" She shoved the teapot over toward Stella who shook her head.

"You pour. I don't want to get pregnant."

Alice laughed. "I wonder where that came from — the shape of the pot? I haven't heard anybody say that in years." She was curious about her new friend.

"Have you ever been pregnant Stella?"

"Hasn't everybody?" She paused. "Listen Alice, I didn't just come up here for a vacation and decide to stay. I mean, I'm broke and it's nice of dad to help out — I knew I could count on him — but the truth is I came up here because I couldn't stand to be in California another minute. Robert — my lover — died."

"An accident?"

"I suppose cancer's a kind of accident isn't it? He died of cancer and he took a long, long time to die. I spent the last week living at the hospital, in an area they call Widows' Walk. Just before he died he asked me to make love to him. So I locked the door and did it. But I think I got some of his death in me — I knew I had to get out of there quick."

Alice looked at Stella's small hand with its nicotine-stained fingers. She wanted to put her hand over Stella's but felt instinctively that Stella wouldn't want that or like that so she just sat there, trying to imagine it all.

"I have to go back down there sometime," Stella said, "to exorcise him. But right now I can't. I tried once, at the end of the summer. I got as far as the Vancouver airport and turned around and came back."

"And Harold?"

"Harold is great. Harold is all energy and life. In a way I guess he's my therapy. I just wish he wasn't so damn noisy. God, the deaf are noisy. And now we're getting a dog."

"Why?"

"Why not. Here we are in the country. Why not a dog? Might as well do the whole thing. *You* have a dog."

"We have kids. I always had a dog when I was a kid although my dog was much better behaved than Byron. Byron's nuts. He was hit by a car when he was a puppy and still he never learns. Even with a limp he can almost outrun a car. He'll chase anything that's moving, including deer. Sometimes I think we should put him down. What kind of dog are you getting?"

"A Saint Bernard puppy," Stella said. "We don't do things by halves."

OCTOBER

Limpets do not go limp — would be no good at civil disobedience. Would not hang limp [3]flaccid, pliant, (2) from *lampen,* to hang loosely down[2] in policemen's arms. No, these little mollusks creep over rocks, feeding on algae, but always returning to the same spot. "The muscular foot is so powerful that limpets are found in wave-swept areas where few other forms of life can survive."

We could all learn a lesson from limpets, dear little conical creatures with Chinese hats. They really know how to hang on.

Which was not really what Alice was supposed to do— or was not really what she was supposed to do re: Peter, her husband, formerly Peter the Rock. None of that hanging on for dear life or limpet life. Rocks nippled with limpets above the pure, bright sea, where she and her daughters crouched, studying the creatures in the intertidal zone.

Stella had a friend visiting her, a woman named Trudl. She and Stella had been friends since high school and now Trudl had left her husband and was looking for a place on the island.

Alice had made a sign which she hung on the fence. NOT OPEN TO THE VISITING PUBLIC BEFORE 3:30 MON—FRI. She explained that she was writing a book. ''Or trying to. Life keeps getting in the way.''

''I'm trying to paint,'' Trudl said. ''Right now I don't know if it's art or therapy.''

''We're going to have painting sessions every morning,'' Stella said. ''The way you keep at it is an inspiration to us all.''

''Well, it gives me something to do with my hands.''

''Where's *your* daddy?'' Trudl's daughter Christobel said to Alice.

''You mean my husband.''

''Yes. Trudl's and my daddy cried when we left, did you know that?''

''All right Christobel,'' Trudl said. She smiled painfully at Alice. Alice smiled back. She wondered what it was like to be the leaver and not the left. The two women liked one another right away.

''I'm glad you're here,'' Alice said.

"I'm glad you're here too."

"Come over any time."

"After three-thirty!"

"Right."

Before they left Trudl showed Alice a small painting of a thin, red-haired creature crawling out of a big cardboard box. It was very badly painted but the message was clear.

"I'm not out yet," Trudl said, "but at least I'm moving."

Alice felt that she and Trudl were being shown off to one another by their mutual friend Stella, and it made her feel a bit awkward. She wondered why Trudl had left her husband. He had cried when she left so it wasn't a joint decision. Trudl had straight red hair and bangs, freckles and a big smile. She looked the opposite of Stella—she looked fresh. The girl next door. But she had a husky, sexy voice, a much sexier voice than Stella's. June Alyson had had a voice like that, however, and had still managed to be the girl next door. No doubt Stella had clued her in about Peter. He was coming out that weekend; Trudl would probably meet him. Alice knew it was good for her to go into town and have a break but she almost wished that Peter didn't have to come out here. It would be nice to have some friends who were exclusively her own. She shivered. Someone had just walked over her grave.

OCTOBER

—Why didn't I tell her that I don't "keep at it." Why didn't I tell her that some days I just write my name over and over, very fast, hoping Hannah won't notice. And that sometimes, if Hannah is out chopping wood or gone for a walk with Stella and Harold and Glenn I just do rows and rows of ovals, like this, the sort of thing we did when we were learning to write. In my present state they look more like the rolled wire they put on the top of prison walls. I envy Stella and Trudl, being able to paint together. I can't imagine writing with somebody else. It always amazed me that Peter could paint or sculpt with the radio on or pick up a pad and start to draw at the breakfast table. What I do is such lonely work—like sitting all day in a dark cave with a single candle. Trying to see, trying to see. When they went back down the path I felt very lonely and wistful and I remembered that old saying from primary school: "Two's company. Three's a crowd." I'm just being paranoid, more so than usual. It was nice of them to visit.

It was Thanksgiving weekend and Peter had come over to help pick apples. Alice had spent the week before keeping busy. None of the girls had asked any questions but they all watched her carefully. She would catch them looking at her. She felt in some awful, deep, fundamental way that she had let them down. But she really didn't want to talk about it. She was afraid that if she started she wouldn't be able to stop. And she remembered her mother going on and on and on with enormous catalogs of grief and grievances, even writing long letters to Alice and her sister on yellow secretarial pads, shoving these letters under the bedroom door. She would not be like that; she mustn't. And she didn't really have any grievances against Peter except the one big grievance that he didn't love her, that he had declared their marriage over and expected her to agree. Outside of the girls, Stella and Harold were the only people who knew. Stella said she found it hard to believe, she had thought them the ideal family and had envied their closeness and the way they all seemed to be having fun.

"Well, I could always make him laugh," Alice said, "but that's not the kind of fun he's looking for."

Peter had phoned twice to see how she was getting on. "And now he's making a house call," she thought.

''I feel as though there's been a death in the family,'' Alice said, after the children were asleep—or at least in bed.

''Can't you look at it the other way? As a new life. For both of us?''

''No, I'm afraid I can't. Do you know how I managed this week?''

''How?'' They were sitting on the rug in front of the pot belly stove. All the lights were off and Alice had lit a candle.

''I chopped a month's supply of kindling. I pretended I was chopping off Anne-Marie's fingers, one by one. Only they grew back, over and over, like something out of a myth. And then I had the pleasure of chopping them off again. You lit the stove this morning with a dozen of Anne-Marie's fingers.''

Peter inhaled deeply, held the smoke, let it out again, hunh, hunh.

''Stop it.''

''Why? It's good to get rid of your anger that way. If I'd had any chickens to spare I would've wrung their little heads right off. If I'd had a pig I would've hung it upside down and butchered it and caught the blood in a brass pan. Drunk it, maybe, pretending it was Anne-Marie's blood. ''Drink this in remembrance.'' Ritual sacrifice. Only all I had was some old bits of cedar. Anne-Marie's fragrant ritual fingers. Like something out of the Song of Songs. ''Make Haste O My Beloved.''

''Please stop.''

''I'd like to, I'd really like to. But my mind keeps spinning off. I wish I could slow it down or shut it off or something.''

''Would you—hunh, hunh—like to go for a little walk?''

''Up on the ridge in the dark, perhaps along one of the forest paths? No thanks. Bad things happen to me on walks. People shoot me down with words: I NEVER LOVED YOU BUT I'LL ALWAYS CARE FOR YOU. It's the hunting season. The forest isn't safe, even at night. I've heard them up there with their guns. You wouldn't want to get shot either; you've got so much to live for.''

Peter took his sleeping bag and the flashlight and went out to the shed to sleep.

"I'm sorry," he said, "I just can't take this."

"You have to take the bitter with the sweet, you know. What did you expect?"

"I don't know; I honestly don't know."

The next afternoon, as they were picking the apples, a strange car drove up the back road. Byron tore around the garden fence barking wildly. Peter called him back. The car stopped and a man and woman and young girl got out. They began to walk toward the path by the shed.

Alice held tight to Byron's collar. "It's all right," she called, "he's really friendly. Just showing off."

The little girl had on white leotards and patent-leather party shoes. Her mother came forward to where Alice and the children were standing under the tree. They had been catching apples and sorting them. Peter was up on the ladder.

"Good afternoon," said the woman. "Lovely afternoon, isn't it?"

"It certainly is," said Alice with false heartiness. The woman's voice had given her away. "You don't get many afternoons as lovely as this one. Have you lost your way? Need the bathroom? — we have one — or the telephone? — we have that too. A drink of water? What can we do for you?" ("Jehovah's Witnesses," she whispered to Hannah. "Just what we need.")

The woman had a thin, sweet, slightly nasal voice.

"Well, we'd like to think we could do something for *you*."

Alice ignored this. "Would the little girl like an apple?"

"Oh, why yes, thank you."

The little girl said thank you to the lady.

The man explained that they were traveling around the beautiful Gulf Islands that afternoon to talk to people about peace.

"How nice," Alice said sweetly. "Now *there's* a subject." Peter came down the ladder.

"I'm going to get a beer," he said, "do you want one?"

This to Alice. He ignored the couple except for a brief nod.

''I think you would all agree,'' said the woman, ''that people around the world are weary of war?''

''Oh yes,'' Alice said, ''oh my yes. I couldn't agree with you more.''

Hannah and Anne, who knew that voice, knew what Alice was up to, were trying not to giggle. Alice suggested they go and baste the turkey and take Flora with them. ''If she eats any more apples,'' Alice said to the woman, one mother to another, ''she'll get diarrhea.''

The man said, ''In the combined arsenals of the United States and Russia there is destructive power equal to five tons of dynamite for every man woman and child on earth today.''

''Probably every dog and cat as well,'' Alice said.

''Is there any evidence, with this kind destructive power available, that peace could come in our time?''

''Oh I doubt it,'' said Alice, who, in the early years of her marriage, had invited everybody in — Mormons, Witnesses, Avon ladies, Watkins men—it was all so fascinating, this world of people who came to strange doors. She knew what the appropriate responses were.

The man nodded. ''The dramatic moves by the United States, China and the Soviet Union toward what the French call *rapprochement*—a coming together in cordial relations—has resulted in a cease-fire. This *in spite of* the ideological differences between the great powers.''

Now it was the woman's turn. This was a carefully orchestrated duet. Would the child perform as well?

''Nineteen hundred years ago Bible prophecy told of a time when men would be proclaiming 'peace and security'! That prophecy seems to be rapidly nearing fulfillment.''

''Yet that peace and security will be short-lived,'' said the man.

Peter had been leaning against the back of the house, drinking his beer. Now he came forward to the group under the tree.

''Why do you do this?'' he said to the man.

"I beg your pardon."

"This. Why do you get all dressed up in your best clothes and run around the islands on a day like this? Why don't you take your shoes and socks off and sit down on the grass and just be."

"Live in the now," Alice said. "Groove on all this sunshine. Eat an apple."

The man looked puzzled and distressed.

"I admit that on a day like today it is difficult to believe in hunger and famine or the wrath of God. But the day of Jehovah will come, as the inspired apostle said, 'like a thief in the night.'"

"Oh we know all about *those*," Alice said, "thieves in the night. Some of my best friends are thieves."

"Perhaps you've found some kind of peace within you," the woman said hesitantly.

"Oh no peace, no peace. Everything's in pieces here. We've already had the day of judgment."

Peter was suddenly furious. He began to shout.

"Are you two at peace? Are you loving toward one another and that kid? Do you laugh together? Do you touch one another? Do you have bodies underneath those clothes?"

Alice was shocked. "Stop it, Peter. That's not fair. I'm beginning to blush for both of us."

The woman swallowed.

"I think I can say we love one another."

The man nodded. How old were they? Young but already old. Hard to tell. "I think I can honestly testify to that."

"Well *good*," Peter said, his voice still trembling. "Now if you'll excuse us we've got work to do." He climbed back up the ladder.

"Good-bye," Alice said. "I apologize for both of us. This is a very strange day for us and so we're acting strangely. I think you had better go. We are in fact estranged and I, for one, know all I want to know about judgment day."

The father, mother and child turned their backs and

quickly walked away. Alice stood under the apple tree, tears rolling down her face.

''Alice,'' Peter called, ''will you shift the piece of foam? There are some gorgeous ones at the top and I don't want them to bruise.''

''I'm crying!'' Alice shouted.

''What?''

''I'm too busy crying!'' There were boxes and boxes of apples. What was she going to do with all these apples? She wiped her eyes on the end of her shirt and did as she was told.

OCTOBER

I have been up to see some of the paintings that Trudl and Stella have been doing. Trudl's are pretty, pastel colours all bleared and smeared, nothing so strong as that first one she showed me, the woman crawling out of the box. Trudl never raises her voice. Of course that is the fashion now (listen to Selene, to Raven, to the soft-spoken Coon Bay people) but, as with Raven, I sense there is enormous anger in there somewhere. Her paintings are false, too pretty, something you might find on hippie greeting cards if such a line were to be taken up by Hallmark. They have no energy. I'm always amused by that Thoreau quote which is used so widely now — marching to a different drummer. I can't imagine this limp crowd marching to any drummer at all. I see them instead (thinking particularly of the Coon Bay people now) as the lotus-eaters in Tennyson's poem: ''on the hills like gods together/careless of mankind.''

Stella has just finished a very strange painting. She bought a bunch of old window shades, green on one side, cream on the other, when she was second-handing. One shade had a stain on it in the vague shape of a cross. Using this, she has painted her dead lover as a modern Christ—it's very disturbing. Even she is disturbed by what she did. Says that soon she must go back to California, even if it's for the last time, and "deal with all that." Her crayon drawing of Harold is tacked up on the wall—all in red. He looks like a satyr or Pan himself. He and Glenn are out jigging for cod while I'm there. Stella urges the *Diaries of Anaïs Nin* upon me, assures me I will love them.

When I go I notice how picturesquely the ax has been thrust into the chopping block, how the chess game is always there on the stump. Stella seems more real than Trudl, but is she? Am I just jealous because they are both "dabblers" and I would like that luxury? Stella asked if I were going to write a book about "all of us." I couldn't decide from her voice whether she wanted that or feared it. She wants me to read her lover's (unpublished) manuscript. He was, she says, a psychedelic anthropologist. I don't know what the hell that means. She showed me a picture of her and him, on his one visit to Vancouver, a few years ago. She is smiling her big smile and sitting on his knee. He is much older than she is, Jewish. "Daddy's little girl."

When I got back Anne was home from school and she and her pal Velma were chopping matchstick kindling—or Velma was showing Anne how to do it without chopping her hand. Hannah and Flora were making baking-powder biscuits and there was a beautiful piece of cod, all skinned and filleted, in the fridge, compliments of Harold and Glenn. I keep telling myself, what more could I want? Well, as one of Hemingway's heroes said, you can get through the days all right—it's the nights that defeat you.

One night in October they all went outside to look at the moon.

"It's *enormous*!"

"It's so orange. Like a pumpkin."

''It's called a harvest moon.''

''Why does it do that mother? Why does it look so big?''

''I don't know. We'll have to try and find out. All I know is that the moon hasn't any light of her own, it's all reflected from the sun. But why it's so big and orange this time of year I just don't know. There so much I don't know.''

''You know a lot.''

''Bits and pieces. Scraps that stick in my head. Nothing whole. We'd better go in, it's chilly.''

Flora whispered over to Alice in the next room. ''I wish I had your bed, then I could see the moon.''

''Do you want to come and sleep on the couch?''

''Yes.''

''Okay. But don't wake the others up.''

She came in dragging her bedclothes.

''Good night mummy.''

''Good night.''

''I want to get in bed with you. I can't see so well from here.''

''Come on then.''

But didn't spend even two minutes in lunar contemplation. Curled backwards into her mother and fell asleep.

OCTOBER

I'm very rational in the daytime, making porridge, making soup, keeping the home fires burning. Night thoughts and poetry begin after the sun goes down. And that too is when I read about voyages of exploration, about the search for the Northwest Passage, about brave bold men in their rickety boats. About Cook and Bligh and Vancouver. About the Spaniards. About Nelson putting his blind eye to the telescope at Trafalgar, ignoring the signals. ''God gave them victory but Nelson died.''

The turn of women, now, to go out exploring? Do we

want to remain like John Donne's mistress, passive — "She is all States/and all Princes, I" — or like the woman in the dress made up of a map, the one we sang about at summer camp and thought we were being so naughty:

Her back was BRAZIL
Her breast was BUNKER HILL
And just a little bit
Below
Was BORNEO

Would we take our children with us, on these voyages of discovery?

FIRST MOM ON THE MOON

Would our lovers wait faithfully for us until we returned? (Would we really want to go?)

It was almost Halloween and Alice bought a nice round pumpkin while she was in town. She decided to carve it and leave it there as a present for Peter. She spread newspapers on the kitchen table and set to work. Peter did wonderful pumpkin faces but if Alice made teeth she usually botched it and they all fell out. This time she was very very careful and began to enjoy herself. She hadn't had a pumpkin all to herself in years. She turned the radio on to the Metropolitan Opera, made a pot of coffee and felt quite content. She would leave it on the dining-room table, no note, none of her awful excessive letters. She wondered if there was a candle end someplace; there was bound to be. Over at the cabin Peter would be carving, or helping to carve, the pumpkins they had grown in the garden. They hadn't come out very big and the girls were disappointed, especially Flora. Maybe she'd make a pumpkin pie with all this pulp. Maybe take the seeds back to dry out and roast.

"Peter Peter pumpkin eater
Had a wife and couldn't keep her"

Only it was the other way around. There was some horrible illustration of a character with a pumpkin head in one of her childhood books. One of the Oz books maybe. How did pumpkins become jack-o'-lanterns anyway? They were really quite scary, grinning their yellow grins from porches and darkened windows. Death's-heads. Worse a few days later, beginning to rot, their mouths all wrinkled and fallen in. She would have to find out where it started. But they were a lovely color, like the sun. She made two not very professional triangles for his eyes and then got daring and added, oh so carefully, crescent eyebrows. He was going to be good. Maybe she would try teeth after all.

She had tuned in in the middle of the act and had no idea what opera it was. Never mind. They were all about love, were they not? A strange art form. Tolstoy mocking it, in one of his essays: "Home we Bring the Brr-eye-id." Alice would rather listen to it than watch it. To watch people singing their passion as though it were conversation put her off. What she really liked was to listen to the music and not even try to visualize what was going on.

Next week all the mothers would be making cookies and decorating them. The supermarkets would have their usual assortment of pumpkin pies and Halloween cakes. When she was little, everybody came to school in costume and, after assembly, where prizes were given for the best in each class, the best in the school (the PTA executive ladies sitting in front and conferring behind their white-gloved hands), each homeroom had a party. She had hated parading across the stage with her class because her costumes were always not quite right. Once she wore a dress made entirely out of *Life* magazine covers. She thought she should have won. She knew her mother wanted her to win, at least for her class. But two other mothers had had the same idea. Her sister, that year, was Johnnie, the midget bellhop who advertised Philip Morris cigarettes. Was her sister's costume bought or made? Hers was good as well and she wasn't shy like Alice. Trotted right up there onto the stage yelling "Call for Phillip Mor-ees." All she could remember was that a girl dressed as the Statue of Liberty had won for her grade and for the school that year. It was probably all horribly competitive, the mothers competing against the mothers, but the homeroom parties had been fun. A cookie that looked like a pumpkin was much nicer to eat than a plain one. Cupcakes with orange icing. Cider in fluted paper cups. Often a Halloween song. What would teachers do without all the annual festivals? Alice remembered bats and witches cut out of black construction paper and pasted on the schoolroom windows. Even the handwriting exercises and the workbooks came back with Halloween seals. Perhaps they could have a Halloween party on the island. Stella and Harold, Trudl, Christobel, Alice, Hannah, Anne and Flora. Lots of old clothes around and bits of ribbon. Hats. A costume party. Harold already looked a bit like something out of Halloween.

Bat, bat
Come under my hat
And I'll give you a slice of ba-con

It was getting dark and her pumpkin was finished. Should she make a pie and leave it? No. Better not. That would be too much. She wrapped up all the pulp and seeds and put them in the bin. Searched in the kitchen cupboards ("Just as untidy as ever," she thought with a certain satisfaction), found a piece of candle and some matches. She fastened the candle to a jar lid with hot wax and placed it inside the pumpkin. Then carried it into the dining room and put it on the big table.

The note had fallen on the floor and that's why she hadn't seen it. Perhaps when she opened the front door a draft had been sent up.

Dear Alice —
The woman wanted more money for the fleece than she originally said. You owe Anne-Marie another seven dollars. I paid her and you can leave a cheque for me. I've taken the wool out with me.

— Peter.

Alice looked at the note. She particularly liked the "dash Peter." He was so careful with the word "love" these days. "Stingy" might be a better word. " —Peter." She went out to the hall, where she'd left her shopping basket, and came back with a bottle of sherry and her checkbook. After she wrote out the check she went and got a glass and poured herself a drink. The newspaper had come and she sat on the couch reading the headlines, the horoscope, "Hang on a bit longer. If you react to competitive challenge you lose out on a shot at worthwhile opportunity." The fillers:

"Hardened paintbrushes can become as good as new if you simmer them in boiling vinegar."

"At 0 degrees F oysters can be stored up to a maximum of three months."

Her mood of contentment had vanished and this in itself depressed her. He could hurt her so easily and so unconsciously for of course, in his new philosophy, the " — " would be automatic. "Love" was a sacred word, to be used only when you really meant it. Maybe he was right; most probably he was right; and certainly he didn't want her to be misled by any false hopes. She poured another drink and then decided to light up the jack-o'-lantern and sit in the dark for a while. Maybe she could get her mood back if she tried. "Switch channels," the way Peter said he used to do when he first started taking dope and sometimes got paranoid.

She sat there and grinned back at the jack-o'-lantern.

"Well my dear, here we are. Did you know there's been a tidal wave off the coast of Japan? A hurricane that's been downgraded to a tropical storm. How humiliating that must be. An assassination in one of the banana republics? A lady in Sweden who gave birth to seven infants? 'They looked like rats,' her husband said. It's just one thing after another is it not?"

She felt like calling up Anne-Marie, ostensibly to thank her for the fleece, really to make her feel guilty. Should she? Why not? It was so easy for her, wasn't it, with Alice and the kids over on the island. Out of sight out of mind. So convenient for Peter too. The second house, the ex-wife safely away in the country. She wondered if Peter had planned the whole thing all along. She had just *thought* the original suggestion was hers.

''Peter, let's try and move over to the island.''

What had he said once, laughing, ''It's easy for you, you can work anywhere.'' He would have to quit his job. That's why *he* wanted to do it gradually, spend another year in town, come out on Thursdays. Or so he said. Actually he could have a kind of captain's paradise going (if he decided to stay with the wife and kids at all) and if he didn't, well, they were all moved out and happily relocated. What would have happened if she had decided to stay in town after all? If they'd only had one house?

''Beggars would ride—'' she said to the jack-o'-lantern.

She went into the kitchen to make a sandwich. Or why not order a pizza all for herself? She'd never had a pizza all for herself. Lots of green peppers and anchovies and mushrooms. A special treat. She looked up Boston Pizza in the yellow pages. It wasn't there. Information said it had changed its name to Olympia.

''Of course,'' Alice said, ''everybody's doing it.'' She wrote the new number down.

Then picked up the phone again and called Anne-Marie.

Her daughter answered. Alice could hear faint sounds of talking in the background. Music.

''Hello.'' Her high thin child's voice.

''Can I speak to Anne-Marie?''

''Who is it?''

''Hi Jeannie, it's Alice.''

Jeannie yelled at her mother.

''It's Alice.''

And then Alice heard Anne-Marie's lovely deep husky voice from across the other room.

"Oh Lord. Jeannie, tell her I'm busy right now and will call her later."

"She'll call you later."

"Tell her never mind."

Alice stared at the telephone receiver. Ugly things. She had always thought so. Purveyors of obscene phone calls and bad news. Also made things far too easy for some people. Peter's new reply if Alice called him from the island.

"Oh. Hello there." "There." An expression like "—Peter." The kind of thing you said when an old acquaintance called you up. Somebody you never expected, or wanted, to hear from again. Who the hell did they think they were? She had feelings, even if she wasn't one of the supersensitive new yoga-and-yogurt crowd. Even if she didn't say "far out" or "rip off" or make all her own clothes. Even if she didn't like Hesse or Kahlil Gibran. They were all so *superior,* these people. She had a good mind to go over there and tell her off, the fucking bitch. Getting her kid to do the dirty work.

A great tidal wave, a hurricane, of hatred swept away Alice and all her common sense. She was running running down Broadway, across intersections, bumping into people with never an I'm sorry, running down Fourth, Third, Second.

One light was on in the little house but nobody came to the door. An uncut pumpkin sat on the stoop. Alice tried the door and found it open but there was no one inside. There were empty wineglasses, two of them, by the fireplace. She was seeing someone else while Peter was over with his children. But of course he would know about it. They wouldn't want to get caught again in that awful "couples" routine. She sat on the couch and stared blankly at the fire, which was almost out. It was a nice house, very small and originally just a summer cottage when Vancouver was still young. But Anne-Marie had made it beautiful. Her batiks hung on the walls and there were huge velvet cushions everywhere. A wicker basket full of yarns sat by the fire. A single white rose, no doubt a present from Peter, stood on the mantelpiece. Alice imagined Anne-Marie's slim, boyish body stretched out in front of the fire. Naked, on velvet cushions, Peter kneeling beside her or in front of her. Alice put some cushions on the floor and lay down. But the room was getting cold and her fit of anger had made her restless. She'd better go. But where? Where? She didn't really feel like going to a movie alone. Maybe she would go and visit Penny and Ron. They were just around the corner. Just to say hello. She liked them, especially Penny, then she'd go home. Sometimes it was good to have a chat with somebody. She was so depressed; she needed something outside herself to cheer her up. Somebody. She'd invite Penny to come over one weekend with her kids. Yes. Out and running again. She rang the door bell and waited.

Ron came to the door.

"Alice. This is a surprise." His north-of England accent sounded homely and reassuring.

"Can I come in? I was out for a walk and saw your light on."

"Of course. There's several people here. We're all in the kitchen eatin' dinner on our laps and pretendin' we like it. Anne-Marie and John are just off to the opera but all the rest of us are stayin'."

Alice began to back away.

"Anne-Marie?"

"Yeah. Come on in."

"I can't. I—there's too many people. I— "

"Have you eaten?"

"No. I was — I was just going home. I'll come some other time." He pulled at her arm.

"Don't be daft. Come on. You're lettin' the cold in."

Alice followed him into the kitchen. It seemed to be full of people she didn't know. Anne-Marie, in a long black dress and black shawl, her hair pinned loosely on top of her head, was sitting beside Jeannie and a man Alice had never seen before.

''Hello!'' Penny said. ''If you'd come five minutes later you'd've got nothing. Grab a plate and we'll all shove over.''

''What do you mean,'' Alice said to Anne-Marie, ''by refusing to talk to me on the phone?''

''I didn't refuse,'' Anne-Marie said, ''I was busy. I said I'd call you later.''

''You told John all about Alice,'' Jeannie said.

''Be quiet, Jeannie.''

''Hello John,'' Alice said. ''You obviously know who I am but I know nothing about you. Are you a husband?''

''I beg your pardon?''

''Are you somebody's husband?''

The room was absolutely still.

''As a matter of fact,'' he said, ''I am.''

''I thought so,'' Alice said. ''Anne-Marie only fancies husbands. She really likes them. She eats them up, don't you, you bitch.''

Then Alice burst into tears and ran into the living room.

OCTOBER

SINGLE?
IN-T-MATES DATING SERVICE (24 HRS)

Peter leaves the newspaper after he's been over here. I never buy it on weekends ''off.'' Mostly I sleep and go to the movies. There's only been three weekends in town so far, anyway, and one of those my disastrous confrontation with Anne-Marie. Disastrous not to her, but to me. Such a naked, *public* exposure of my anger and grief, right out of character. Not that I'm sorry. I was reading a legend about the hyrax the other day, that little big-eyed creature who lives in West Africa and screams all night from the top of a tree. Apparently he insulted one of the largest animals and had to say he was sorry. So he started up the tree crying, ''I'm sorry, I'm sorry,

I'm sorry,'' but by the time he reached the top (and knew he was safe) he was screaming ''I'M NOT SORRY, I'M NOT SORRY, I'M NOT.'' And that's what he screams all night in the African bush.

Over here I read the old newspaper from cover to cover; the front page is not so interesting to me now as the personals.

HAPPY BIRTHDAY
GAYLENE
LOVE, WILLIE

SO LONG POPPY
THANKS FOR THE MEMORIES

HAPPY FIFTIETH
JOANNE AND CHARLES

And the In Memoriams. ''Suddenly, in his forty-ninth year.'' Often ''survived by his loving wife'' and then the names of the children. The wives put them in, I guess. Here's a nice one. ''Passed away in Hope on November fifteenth.'' Survived by his wife. Etcetera. No ''loving wife.'' Perhaps she was glad to get rid of him.

MOTHER FEARED WOLVES
GIRL DIES AFTER FALL

MURDERER SET WIFE ADRIFT ON RAFT

PARK HID BODY
FOR FIVE MONTHS

BOGUS NURSE STEALS BABY

And then I look up a quote from Huxley. ''One is always alone in suffering. The fact is distressing when one happens to be the sufferer, but it makes pleasure possible for the rest of the world. *Crome Yellow.*'' I wrote that on a card years ago and kept it. Can't remember why.

Here's a man who caught a shark and inside was a human leg, complete with sock and shoe.

''It just popped out,'' he said.

The nightingale is one who yells at night.

The only way Alice could be alone was to go for a long walk or get in the car and, on some excuse, drive down to the south end and back. She found it difficult to yell and drive, but voices carried in the woods. Once a month she went into town, but there was the tenant in the basement suite — there were the girls upstairs — they might come running. Once a month the children went into town and that was the time, pounding the floor, weeping, rage rage rage. The cat would remove herself to the top bunk in the children's room and the dog would sit by Alice, worried, sometimes thumping his tail, thinking it all a game, sometimes licking her hands, her arm, with little anxious licks. Joni Mitchell assured her that you don't know what you've got till it's gone, Joan Baez told her the answer was blowing in the wind, the Beatles said to let it be, let it be, let it be. Peter's pipe was on the mantelpiece; his plaid shirt hung on a hook by the front door. She slept alone in the big bed in the new room, grateful to have the children looked after for a few days but so lonely, so alone. The golden rug on the floor mocked her. They had had great difficulty deciding on a color. Alice had thought moss green; Peter had felt that a pattern of some sort might be better, what with muddy feet, knocked-over wineglasses, things like that. He had gone into town several times to try and find something suitable. Finally called her one Saturday morning to say he'd seen a lovely gold. He came back with the rug, which was perfect, and a present for her, a poster of Sarah Bernhardt playing Hamlet.

Peter, who had always covered up, now wore bare feet and sandals. His new knowledge of his body made him fearless. She had not been able to do that for him—it was her great sin against him—that she could not release him from his chains. Rubbing the marks on wrists and ankles, deep welts of fourteen years, he stared at her (or so she thought) with the dark and hate-filled stare of the escaped prisoner.

Pudendum, n. (usu. in pl. - *da*)
Privy parts --. (L. *pudere,* to be ashamed.)

Pregnant with Hannah, horny, living in his parents' house, Peter whispering, "Are you sure we should be doing this?"

NOVEMBER

"I wasn't tender."
"I wasn't gentle."
"I took him for granted."
"I forgot to tell him I loved him."
"I assumed, because he was so strong, that he had no secret pain."
"I didn't touch him enough with the tips of my fingers."
"I finished his sentences for him."

"I wish he would come back."

"Alice, there are no victims."

He had been a used-car salesman in New Westminster and his old lady had been a topless go-go dancer and now had something terribly wrong with her feet. It had been raining so Alice invited them up for tea.

"How long were you married, Alice?"

"Almost fourteen years."

"Fourteen years is too long to be married to the same guy."

Alice stared at her cup.

"But he wasn't the same guy for fourteen years."

Alice sat in the bathtub, steam from her body rising like mist in the cold room, the candle flickering from the drafts. She was listening to the six-o'clock news. The announcer said something about "striking office workers." She saw them hitting each other — hard — with sharp sticks. Purple bruises, color of starfish, color of sea anemones, blooming. Blood everywhere.

Peter had given her a black eye. She had gone into town on the Friday night, with the girls. Because she needed a break. Because she wanted to buy flannel and lace for some nightgowns she was making as Christmas presents. Alice was all right with sewing so long as there were no darts, although she sometimes had a little trouble with sleeves, fitting them nicely into the shoulders. Years ago, when she was first married and they were living with Peter's parents, his mother had asked when she was going to start knitting for the baby.

"I don't know how to knit," Alice said. Peter's mother was shocked. Alice might just as well have said she didn't have breasts or a vagina.

"I'm willing to learn," she added weakly.

The next day Mother had gone to the village and bought wool and patterns. Blue wool and white, soft colors like a summer sky. The lessons began.

"I like the *idea* of knitting," Alice said to Peter a week later, "it's my hands that don't." Peter could knit; he had learned at art school. Peter, like his mother and father, could do just about anything with his hands. They had a phrase for people like Alice — "cack-handed." She never heard anyone use that phrase except his family.

Alice persevered, swallowed pride and went to Mother when stitches dropped or wool tangled. She made one small blue matinee coat, complete with buttons. She embroidered simple flowers on one small Clydella nightgown. But the word had got out — to the grandaunts, to the ladies at the bridge club. Alice could not knit. Hannah was inundated with matinee coats and nightgowns and booties. Alice and Mother gave a tea and showed the baby off. Alice's hands discovered they were pretty good at a lot of things, that their lack of precision disappeared when bathing a child or soothing it. That they knew a lot about stroking.

(Something that they had tried to convey to Peter, but he, in bed with a woman so obviously pregnant, would whisper, "Do you think we should be doing this?" Or, if they did it, Mother would call from the next room, where she and Father slept in easy familiarity in their twin beds, night table and lamp between them — "Is everything all right in there, dear, I thought I heard a noise.")

Alice went into town that day for a lot of reasons. She had not intended to stay overnight — the girls were spending the weekend with Peter and she was going back to the island to have some time on her own — but she had a bad cold and he suggested she stay over. Flora had a bad cold too and took a long time going to sleep. When it was all quiet down in the children's room, Peter lit the fire and brought out a bottle of wine. Alice looked around the sitting room. On the mantelpiece there was a single red rose in a glass; the water had dried up and the rose was so old it was nearly black.

They sat in front of the fire, sipping wine and saying nothing, but it was not an uncomfortable silence. Then Peter spoke:

"Do you realize this is the first time in a long time you haven't tried to lay any trips on me?" Alice wanted to say,

"I hate that kind of talk" but for once she said nothing. She picked up the octoscope and began watching the fire. After a while, after he had toked up and sipped a bit more wine, Peter spoke again.

"I can make up a bed for you on the couch and I don't mind driving you to the ferry in the morning. I'm going to bed now and if you'd like to join me, you may."

Alice's nose was running.

"My nose is runny," she said.

"That's all right."

After they made love Peter said, "That was nice," and went to sleep. Alice lay awake, her hand between his legs, knowing she must not say "I love you," knowing that would spoil it all. Knowing she must not say, "Please, I want you back." Saying over and over in her head, like a rosary, all the things she must not say.

Because she was so wide awake she heard the phone ring first. She started to get up, quietly so as not to wake Peter, then, remembering that this was no longer her house, or rather, no longer the place where she lived, she lay back down and let it ring. It was very late, very very late. Who would be so insistent at such an hour? On the eleventh ring Peter came awake and went to answer the phone. Alice could hear him talking softly in the kitchen.

When he came back into the bedroom he began to dress. Alice pretended she had just woken up.

"Peter?"

"I have to go out," he said.

"Who is it? What's the matter?" He came over to the bed, tucked in his shirt and leaned down.

"Is it Anne-Marie?"

He put his hand over her mouth, "Don't ask." He went out and shut the door behind him.

Alice got up and stripped the bed, rolled up the sheets

and put them in the corner. Then she went down the hall to the bathroom and began running a bath. She took new sheets from the cupboard, went back to the bedroom, turned on the light and made the bed, squaring the corners, pulling everything tight so that there would be no wrinkles. She took her clothes with her into the bathroom and sat in the bath for a long time. After she had dried herself she scrubbed out the tub, dressed and went to the sitting room where she built up the fire again. There was a bit of wine left so she drank it. She could hear Flora coughing in her sleep and wondered if it would be possible to leave before Peter got back or if Flora would wake up and be scared. What time did the bus depot open? At what hour could she go and sit there with all the drunks?

Peter let himself in quietly and seemed surprised to see her up and dressed and sitting by the fire. He looked tired.

"I would like you to take me to the ferry now," Alice said, "or if you don't feel like that, I'd like you to take me to the depot."

Peter sighed. "Why can't we all love one another?"

"Let me know when you figure that one out," Alice said. "Meanwhile I want to go to the ferry."

"The ferry isn't for hours."

"I don't care. I don't want to stay here. If you don't want to take me I'll call a cab. That might be better anyway, Flora's been coughing a lot."

"No," he said. "I'll take you, you're upset. I'll leave a note for Hannah."

In the car Alice said, "Did Anne-Marie know I was coming in today?"

"I expect I mentioned it."

"She is very clever," Alice said. "She knows you inside and out. Did you know 'manipulate' means to hold in your hand?"

"She couldn't breathe," Peter said, "she has these awful attacks where she can't breathe and they frighten her."

"When was the last one?" Alice said.

"I don't remember. She seemed to be getting better."

"And what do you do when she gets these attacks? How do you help her? Do you go down on her or what?"

"I just hold her until she goes to sleep."

"I once saw a movie," Alice said coldly, "called *The Egg and I*. I think it was Claudette Colbert and Fred MacMurray. You probably never saw it. Did you see it?"

"I don't think so."

"No. It probably never made it across the water. Anyway, they run an egg farm and a female journalist comes out to interview them—I forget why—and starts making up to old Fred. Claudette realizes what's going on and says, 'Excuse me, I've got a headache,' thus putting an end to the interview.

"The journalist says, 'Do you get these headaches often?' and Claudette says, 'Often enough.'

"I've always remembered that line. Don't you think that's a funny line?"

"Very funny."

"So I think it was very convenient that she had an attack tonight," Alice said. "I think 'attack' is the right word. I think she planned her 'attack' very carefully—to find out whether I was still there. I think she's a fucking bitch." Alice began to shout, "Fucking bitch, fucking bitch, fucking bitch" at the top of her voice.

That's when he hit her—hard—and then, turning onto a side street, he began to cry.

When the announcer said "striking office workers," Alice saw bruises blooming, purple as starfish, purple as her left eye.

The next night the announcer mentioned a man who had "murried" his wife. Then he said, "Excuse me I meant murdered and then buried." Alice laughed. She often talked back to announcers. "Sounded like married and murdered to me," she said. She wondered if the announcer were having problems of his own.

Alice had smoked some of his precious dope in the middle of the week after Peter punched her in the eye. Sitting on the front porch she watched Flora, in her red sweater, playing with all her babies. The child's face began to change, was unbearably beautiful and then too bright, it glowed, Alice couldn't bear to look at it, and then it was the face of an old peasant woman in a scarf and an old red sweater. Alice was fascinated and terrified at the same time. In her distress she got up and walked down to the wharf. She sat for a long time with her head on her knees, trying not to think of anything, look at anything, afraid to go back up and confront the old dwarf woman on the porch.

Peter came to visit her and brought her African violets.

"When I finally went up," she told him, clutching the little pot of flowers, "Flora was just Flora again, her fat round lovely self." Alice's eyes were full of tears. "It was scary."

"Why scary?"

"I don't know. I seem to have enough trouble with reality as it is."

"Don't give in to fear."

"That's easy to say."

"Switch channels if things start to get too much. I remember when I first started smoking. I had a few bad moments. I just learned to concentrate on something else."

"But what did it mean, Flora glowing, and then an old crone?"

He shrugged. "Don't try to analyze it. Just accept it. You were projecting fears of yours onto the child. Your own fear of death, perhaps."

He sat on the couch in the new room. She wanted to go to him, but couldn't. Could not. Was not allowed. It was as though he had drawn a magic circle around himself and she could not set foot across it no matter how hard she tried. If the children thought it strange that daddy was sleeping on the couch they said nothing.

"I've never had much luck with violets," Alice said. "I think they're rather delicate."

NOVEMBER

I had dressed for the occasion (my purple sweater, my paisley skirt) and even had arranged some deep purple hydrangeas in a pot by the bed. They were almost exactly the same color as my eye. After he left, still in my "costume" I went out and chopped up kindling with the red-handled hatchet. I did a whole box, slim as chopsticks and brought it in in triumph.

"That should keep us warm."

I was anxious for nightfall and a fire in the big room so that I could begin burning Anne-Marie's beautiful long fingers.

"Bertrand Russell said he didn't think there would be any more wars," I told Hannah, who was sitting studying her French at the kitchen table, "if there were a chopping block in every backyard."

"What about any more trees," said Hannah, without looking up. She was in the kitchen the whole time we were having our "conversation." Peter says that when he hit me he realized how much he did love me (how nice) and that he has decided to find a little basement suite somewhere so I don't have to be "disturbed" if I want to come into town. I just listened to it all and then he went away, God knows where, maybe to visit Stella and Harold or maybe just to sit in his car until the ferry came.

I am so torn! If I went to a lawyer now I could sue for divorce and have it over with, done with, get on with my life. I could even take the girls away someplace, for a year or so until the wounds start to heal. I could ask him for the money to do that, *demand* it. And yet a part of me is so convinced that he'll come back, that I must hang on, stay here, not move the girls around for a while, finish my book. But I understand, now, about wanting to make a man pay and pay and pay. I understand about vengeance.

Stella says it's worse if your lover is dead but I don't think so. At least he isn't calling up saying "hello old friend" or punching you in the face if you dare tell him he's a hypocrite and a jerk.

The girls try not to look at me, the mark of their gentle father's anger blooming on my face. Well, it will fade. On the outside anyway, on the outside.

I suppose it is hard, maybe impossible, to admit to the wife of your youth that you want Romance, you want somebody to see you as mysterious, worthy of being explored. That you want to be "new" to someone, that what you have with one another is no longer enough. He doesn't even want to talk about it; he doesn't even want to try. He said one day that he realized he had never been in love before (Anne-Marie), that perhaps one only falls in love like that once. I was going to go and get his letters to me, the letters he wrote before we were married, and show him how wrong he was, that "in love" is intense and transient but that what follows is equally good and perhaps more interesting. But I could not humiliate myself that way; he would have had some rationalization. "Oh I *thought* I was in love with you . . ." some such thing.

Even old Horatio Nelson doesn't help me tonight. "This is too warm work, Hardy, to last."

My dreams are terrible lately. Sometimes just undifferentiated terror is all I remember as I force myself awake, just a tidal wave of blackness. Sometimes Peter is in them and we are making love. In one I was stuffing enormous, uncooked turkey necks into a green garbage bag, ostensibly to take them up to Stella and Harold's for Uggah their dog. Thank you doctor, I can figure that one out for myself.

Everyone had dressed for Alice's birthday party. Harold had come as Van Gogh and he was the winner. Stella wore a fishnet for a veil. They all sat around the pot belly stove, a little tipsy from the birthday wine, reluctant to break up the party.

"I wonder if Durrell's friends were all just as ordinary as we are," Stella said, "and he just touched them with his magic wand?"

"D'you think we need a Durrell or a Prospero then, to enchant or immortalize us?"

"Are you going to try it, Alice?"

She shrugged. "I don't know. After I finish this book I don't know what I'll do. I'm not sure I want to transform us all into characters. And perhaps I am still hoping for the, oh so unacceptable in literature, happy ending?"

NOVEMBER

Reading "The Princess and the Pea" to Flora tonight. "There once was a Prince who sought him a wife." Even princesses have to pass the test, this one very peculiar — if you're covered in bruises you're sensitive enough to marry the prince. And the old mother cheering him on. Peter dotting my eye then coming to me and saying, "When I hit you, I knew I loved you." Perhaps I, too, bruise easily?

I wonder if the prince tested her once a year, just to make sure she hadn't toughened up as she got older?

Selene arrived at the door one day, with her backpack.

"Raven will probably show up tomorrow or the next day, but I told him I wanted to talk to you first." Her voice trembled.

Alice put the kettle on and she and Hannah stopped working.

"What's the matter?"

"Things haven't been going too well for us." She shook her head. "It's probably my fault, but I feel as though I'm the only one who cares if things are clean or tidy. Raven and the others just do not care. They *expect* me to do it all. The latest thing is cat shit."

"Cat shit?"

"We've got this kitten—it sort of belongs to all of us but it spends most of its time in our tepee. It shits everywhere. Nobody seems to care but me. I just had to get away." She began to cry.

Alice took a deep breath. She remembered only too well what Peter had said about her "bloody book" being more important than her friends. How would it be if Selene stayed here? She was very quiet but it would be one more body in the house, one more person to give to. Alice was tired and the work wasn't going all that well. Yesterday morning Trudl and Christobel had come by too early; Alice told them they'd have to come back around three. It embarrassed her to say that. Trudl had said she understood but Alice wasn't sure.

Before anyone saw the sign they could see her, see Hannah, when they got to the top of the path. If she put her sign at the bottom of the path? What pretentiousness! One of the older men at Coon Bay, a sixty-year-old ex-NASA type, born-again hippie, had told Alice that you couldn't legislate creativity. She should always be open to whatever presented itself on any particular day.

And now Selene, whom she really did like, who needed help. Alice pushed down her resentment.

"You can stay here as long as you like."

"I know that," Selene said, "but I think I want to go to New York to see my mother. Really get away and think about it all. I was wondering if I could phone her?" Alice took another deep breath. Selene had "known" she could come and stay, even though she also knew Alice was struggling with a book.

''You'll have to call collect.''

''Oh, of course. I'll wait until after six.''

''I'm not sure it makes any difference if you're calling collect.''

So she called her mother at her office and of course it was all right, come right along, she'd send the money for the fare.

''I want to go by train,'' Selene said, ''I don't want to fly. It's not the sort of thing I want to do to my body.''

Her mother would send her the train fare. Right away. Wire it to Peter's bank account in town. Selene would stay with Alice until Raven arrived and they talked, then she'd go into town and get the money, buy her ticket and get on the train.

Selene called Peter. He was delighted to help out. If she came on an evening ferry he'd pick her up. In an hour it had all been arranged.

Then they had lunch. Then Selene and Hannah went for a walk so that Alice could have a little time on her own. But it was 2:15 so Alice just sat and did nothing at all.

Except feel a fierce resentment that everybody seemed to have somebody to lean on except her. That wasn't fair, of course, and as usual she was ashamed of the way she felt and decided to make a vegetable curry, a really nice one, for supper.

Selene was sitting on the big bed, propped up on pillows, trying to catch her breath. Alice was rubbing her back.

Raven came in with some soup in one of Alice's bright orange bowls.

Selene turned away.

''I thought you said you were hungry,'' he said, holding out the bowl.

''I can't eat anything you offer me Raven, I can't.'' Her words fluttered out of her mouth like something mechanical, something running down. Alice wondered if she should try and get her over to the hospital on Saltspring Island.

"I *love* you, Selene," he said.

"Go away, Raven. Just go away." She began to cough again. The children, frightened, sat quiet in the other room.

Selene lay rolled up in blankets in front of the fireplace. She was still having trouble breathing. Alice wondered how she could ever have been afraid of her. All Eve's sisters bobbing for apples. All connected like some vast archipelago. Herself. Selene. Stella. Trudl. The girl children. Connected by femaleness and by blood and by the moon. Yet can't do without Adam.

"hand in hand with wand'ring steps and slow — "

And happy women have no histories.

NOVEMBER

To escape is to take your cape off. But presumably someone is left holding it, while you dash away unencumbered. Selene has gone. Raven has decided to stay here for a while. So now we are five. He sleeps at the foot of my bed, near the stove. I asked him if he wouldn't be more comfortable (that is have more privacy) out in the shed. He said he wanted to be with us. I write this while he sits a few yards from me, facing a candle and meditating. One of the "lost boys" I thought to myself tonight. I don't see how I can turn him away. And maybe it will be good to have somebody here for a while — as "leavening." He lets Flora crawl all over him, is teaching Hannah and Anne to tie knots and make buttons from juniper wood. I just wish I had one small space for myself — a cupboard would do. Maybe *I* should sleep in the shed?

"Good night Raven."
"Good night little sister, sleep well."

Up and down the island, singing as they went

I'll build a bamboo bungalow for two
Bungalow for two MY HONEY
Bungalow for two
Walla Walla Walla

And when we're mare-reed
Happy we'll be

Under the bam-boo
Under the bam-boo tree

Raven told them that the freaks at Coon Bay called it "the house of beautiful women." They (or Alice and Stella—Hannah and Anne thought it was mean) had names for the people at Coon Bay too but somewhat less kind. Alice and Stella and Trudl sat on the steps of the store and watched them all come out of the woods on mail days. A lot of them had dogs and the dogs sniffed and shit and fought (Byron right in there, essentially a hippie dog, doing his own thing) while their owners went in the store for milk or apples and made disparaging remarks about all the "chemicals" on the shelves. They waited for their unemployment checks or welfare checks and their parcels from home. They got letters in handmade envelopes, names and addresses printed in big block letters. To Alice it seemed like summer camp. She imagined parents in Arlington, West Virginia or Titusville, Pennsylvania (the majority of them were from the States) saying over drinks and dinner—

"Oh yes, Marilyn and George are living on this cute little island up in British Columbia, Canada. We sent them off a package today. It's just like Adam and Eve in the garden, my dear. We're thinking of going up for a visit.''

Sometimes they would do just that. Arrive on the island in their big cars, knock on Alice's door if the store were closed, wanting to know the way to Coon Bay. The mothers in white trousers and smart blouses, the fathers in classy slacks and sports shirts, everything about them saying ''money'' and ''America'' and ''middle class.'' Sometimes Alice invited them in while she drew a map. They thought her old cottage darling and were surprised that she lived out here all alone with her three daughters. They asked if she knew Marilyn or George or Dwight or Penny. She hesitated and asked for a description. Penny was Shula now, Marilyn was Blackberry. She did not tell them this — not out of some desire that they be surprised when they got over there, but because she thought she might laugh. And she did not want to be so obviously aligned with these good clean people who were so clearly puzzled by it all. Alice was puzzled too. And contemptuous. She did not say that when George and Marilyn or Dwight and Penny walked out of the woods in their overalls and gum boots, their long skirts and shawls, she and her friend Stella and whoever they were sitting with on the steps of the general store, would say, in mockery,

''Here come the Peters with their Wendys.'' ''Here comes the yoga-and-yogurt crowd.''

If it were dark she let the Mothers and Fathers use her telephone to call back down to the Lodge to see if there was a room available for the night.

''And you really cook on this stove?'' the Mothers would say, marveling. ''My grandmother had a stove like that.''

The Fathers would start looking at their watches.

Once, when Raven was staying with them and Alice and Hannah and Anne were out for a moonlit walk, he answered the door stark naked and there stood a Mother and Father.

Alice arrived back just at the end of the interview. Raven very gracious, finishing up an elaborate map, leaning over the kitchen table with his white bum sticking out, reassuring them that everything was cool, their son was fine. The

Father was so red he looked as though he might have a stroke. They fled down the path but not without a shocked glance at Alice and the two young girls.

"You really are naughty," Alice said, trying to be stern. "I have enough trouble with your nakedness."

"I thought it was you guys," Raven said, smiling. "I'd been meditating in front of the fire."

"Do we knock before we come into our own house? Sometimes I think you're just as big a jerk as all the rest of them."

Alice wondered if Father, in preparation for the day to come, would get smashed in the bar of the Lodge, tell his tale to the barmaid and whoever would listen. Alice had become used, if not immune, to the gossip and speculation about herself, but she didn't want any gossip about her daughters. When Raven left (which she tried not to think about) she didn't want a bunch of drunken yahoos descending on the place, having heard that tale and thinking she was free and easy. Sometimes she saw a kind of quiet aggressiveness about Raven and Selene and their friends. It was in their voices, their philosophy. Anger had been outlawed. Jealousy. Suspicion. Fear. Anybody who got angry, jealous, suspicious, who even *raised their voice,* was somehow inferior. Alice remembered a few months back when Selene was staying at their house in town with Raven. Alice and the girls were bringing a huge load of library books into town and Peter had offered to pick them up at the ferry. On his way to meet them the car had broken down and he sent a message to the ship. Alice's name came over the loudspeaker and she was called to the bridge. The trip home had been awkward with the wheeled cart full of books. The city bus driver wasn't sure he was going to let them take it on the bus. The girls had their packs as well because they were spending the weekend.

Selene opened the door and they all began to talk at once, about the car breaking down, about the loudspeaker, everything. Everybody was very excited. Selene put her finger to her lips and said "Shh." Alice wanted to ask what right she had to come to the door of *her* house and say "shh" to *her* children. They did that, she thought, they had a nice quiet way of taking over. Selene had asthma attacks, bad ones, every time she got a letter from her mother but she told Alice she had gone beyond the "first warrior" so she had at least got beyond fear. Alice was intimidated by Selene and Raven although she knew in her heart they were just as fucked up as she was, and also just as nice. But they weren't nicer — they really weren't more enlightened. Yearned for gurus — always male, preferably from someplace older, more ancient in its wisdom than raw North America. They, too, wanted to be *told what to do*. Or maybe what they wanted was both: to "do their own thing" and to be told what their "own thing" was?

"Selene learned to bake bread from a witch," Anne said one day.

"I didn't know witches ate bread," Alice said, "I thought they ate plump young children who came nibbling at their houses."

"You are so *cynical*," Peter said. "Why are you so cynical?"

"I just don't see why a witch's recipe for *bread* would be superior to any other kind. Unless it was like Alice B. Toklas's brownies. Unless there was a secret ingredient. What if I said I'd learned to make apple pie from the prima ballerina of the Royal Ballet? Sounds like one of those awful recipe books — 'Favorite Soups of the Stars.'" She knew she was going too far. Selene was special to all of them, including Alice. But this hippie bullshit made her mad. She had heard Selene say to Raven, when things were going very badly for them, when it looked as though their relationship might be over, just before she got tired of all the hurt and took herself off for the visit with her mother in New York, Alice had heard her say, very low, "All I want to do is follow you around like a dog." Selene didn't have any more answers than Alice did: she just looked and sounded and *acted* as though she did.

She knew what Peter's answer to that would be. ''Well at least she's *trying* to find answers. At least she's open to it all. You don't believe in even trying.''

Selene said once that she would like to have just one garment — ''one seamless garment'' that would do for all occasions. Alice saw that as a metaphor for what Selene was trying to do with her life. It was a nice ideal, to try and make of one's life one seamless garment, but perhaps only someone like Gandhi could do it. And he had a cause greater than himself, much much greater — the unity of India, Home Rule. People like Selene and Raven seemed to Alice to have given up on the world. They wanted to go back to a paradise before the fall. There was something very romantic, very narcissistic in their attitude. They were nicer than most of the people at Coon Bay but only in certain ways and only because they were more intelligent. The Coon Bay freaks sat in Alice's kitchen and dared her to move them or excite them. They had absolutely no sense of humor, none. They were Raven and Selene carried to the nth power. They were ''takers'' — that was the word for it. Yes. They were takers. They did not care to change the world or make it better (although they thought Dylan and the Beatles were ''far out''); they just wanted to be left alone. Their favorite position was sitting down.

They were all about ten years younger than she was and yet they seemed older, defeated, like people who had been in prison for a long time or deprived of some essential food. It was as though they had had the moral equivalent of a stroke. This was not true of all of them but enough to make Alice uneasy and a little frightened. Had the world done this to them, then? Perhaps ''hope'' implied some belief in the future. Perhaps one had to ''live in the now'' because there might not be a ''then.'' Alice pulled down books from the shelves — poetry, Plato, whatever she thought might rouse them. They glanced at the books and handed them back with an air of polite disinterest. They weren't ''into'' reading just now, some other time.

Alice thought of her namesake:

"'I see nothing on the road,' said Alice."

"'Oh,' said the King, 'I wish I had *your* eyes.'"

When Selene went into town on the first lap of her journey to New York, Hannah went with her to have a break. Later she told of how they sat, one on either side of the fireplace and practiced signing.

"Dad got very annoyed because he couldn't understand what we were saying."

Alice wanted to ask,

"Did daddy sleep with Selene? To comfort her?" but she couldn't do it. Not even in signs. She didn't want to know.

NOVEMBER

"Listen to this," Hannah said. "They want me to make an *apron* for my Home-Ec project. I don't want to make an apron for God's sake."

I said to her, "Why not make a really gross one, all frilly and dear-little-housewife? Maybe you could embroider 'Queen of the Kitchen' on the pocket. Go look and see what we've got in the bottom drawer."

We found some green netting, some black velvet ribbon

and some sequins. We have decided to give it to Stella after it has been handed in for credit.

The girls had gone to spend the weekend with Peter, and Alice and Raven were looking through her old *Saturday Evening Posts*.

"The kids think all these pictures of families sitting around a fancy radio are pretty quaint. But we used to do that," Alice said. "We never had one of those gorgeous models and never had a radio-phonograph console but there were certain things we all listened to. Kate Smith, Amos 'n Andy, Fibber Magee and Molly, the Jack Benny Show, the Shadow, Lux Radio Theater. That was fun. And if we were sick, our mother brought the radio upstairs and we lay in bed and listened to the soap operas. My favorite was Our Gal Sunday, a girl from a little mining town in the west who married England's wealthiest and most handsome bachelor, Lord Henry Brinthrop. They lived in a place called Black Swan Hall and had all kinds of troubles. I particularly liked it on Fridays when the announcer, who had one of those real, rich announcer voices, would say, 'Listen to Our Gal Sunday—Monday.' I've always loved radio; sometimes I think of my dreams as radio programs—but very surreal."

"Hats," Raven said, "all these dudes wearing hats. Far out."

"The women too. My mother always put a hat on if she went downtown. Always. And even when she was leaving us—and she was forever doing that—she put her hat on, and pushed hatpins into it before she slammed the door. Respectability! I don't suppose women in the country wore hats, except to church and weddings and funerals. Did your mother ever wear a hat?"

"I can't remember ever seeing her in one. We didn't go to church and I can't remember any funerals. Maybe I didn't go."

"I wonder why hats went out. The queen always wears a hat—all the royal ladies. But you don't see women walking down the street wearing hats. Fashion is weird."

"You lose a lot of body heat through your head," Raven said, "it makes sense to wear a hat, or a cap in winter." He pushed his fingers through her hair. "Although your hair's so thick you probably have enough insulation."

"I always used to take my cap off even when it was really cold, the middle of winter. As soon as I got out of sight of the house."

"You were probably just showing off your pretty hair." He pushed her hair up from the back of her neck.

"If you keep doing that," Alice said, "I won't be able to concentrate." She picked another magazine off the pile. "The mice and the damp got most of them," she said. "They were in one of the sheds. But there's quite a number from the war. It's sad, isn't it, that we call it 'the war' or people my age do. As though there were no wars before, or, sadder still, after. I think a lot about Peter, growing up in England during all that. Lying in bed and hearing those rockets that cut out before they fell to earth—I think they were called doodlebugs. Waiting for the explosion. And his father away so much. His mother told me that once, when his father was home on leave, Peter said to her, 'When is that man going?'" She pointed to an ad about V-Mail. Picture of proud mother and father, gray-haired, and above them, in the main picture their son shooting down a Jap Zero.

"I saw stuff like that again and again, when I was a kid. There was 'the enemy' and there was 'us.' Everything seemed so simple — or that did at any rate. I was ashamed that my father was too old to join up. That I didn't have a single relative fighting except some remote second cousin. Now I'd be ashamed if anybody in my family didn't refuse to go to war." She gestured at the magazines. "I'll tell you how indoctrinated I was — at home, at school, by all the stuff I saw and read. When I was pregnant with Anne I had a car accident. I was driving along the King George Highway in Surrey and had just signaled to turn into the road where we were living when a great big silver truck, one of those deals with the articulated cabs, passed me on the right, or tried to, not realizing there was a ditch. He swung too close and hit me; the cab jackknifed and he hit me again. We both went into the ditch. I sat in the car, which was partly on its side, utterly shocked, not really hurt but unable to move. The driver of the truck rushed over and started shouting, 'Anybody hurt, anybody hurt.' I looked up at him, and saw he was Japanese; 'Oh,' I thought, 'that explains it.' I thought, in my shocked state, that I'd been dive-bombed by a silvery Japanese plane.''

Raven whistled. ''Those old fears go really deep.''

''And those old prejudices. Same thing I guess. I hope my kids won't grow up with all that baggage.''

''They won't. Your kids are growing up just fine.''

''I don't know if they are or not. I'm so sad about Peter. I try not to let them see it but it's there. Sad and angry and hurt.''

''Don't be angry. Doesn't do any good. You just waste your good energy that way.''

Alice got up off the rug and went over to a chest where she kept paper, paints, glue, scissors and a folder of things she clipped from magazines and newspapers. ''I want to show you something,'' she said. ''I'm going to put it up in the bathroom. It's very funny but it's not.''

She came back with a torn-out page from an old *Post* and handed it to him. ''This is how women viewed men in 1937.''

It was a full-page ad for Community Plate. A Joan Crawford type, perhaps just a little softer and certainly quite young, was leaning out past an elaborate candelabra. She wore a yellow dinner dress and orchids in her upswept hair.

''Jim,'' she was saying, ''this is a moment we'll always remember . . . . Doesn't the candlelight do things? How impressive you look . . . almost as handsome as our beautiful silverware . . . .''

''Notice,'' said Alice, ''that Jim isn't even in the picture.'' She sighed. ''I don't know Raven, do men and women have to go around exploiting one another? I like to think that you people, ten years younger than Peter and I, are doing a better job than we did or our parents did. But is self-fulfillment really the answer? And *can* two people live together in equality? Is that possible? You people never raise your voices to one another — or I used to think you didn't. Peter still thinks you don't."

Raven smiled. "You've heard me raise my voice."

"Yes. Against Selene. Judging her, putting her down. That's one reason she went away."

"I know."

He looked, all of a sudden, very vulnerable. She put her hand on his head.

"You also know she'll come back. She loves you."

"I know that too."

"I wish I knew Peter would come back."

Raven smiled. "Maybe Peter doesn't even know." He stood up. "Let's take a walk. Let's stop all this talking and do some walking. Come on. Byron too." The dog was ready before they were, jumping up against the door.

"It's getting colder again," Alice said. "Maybe it will snow for Christmas."

When they got to the bottom of the path Raven put his arm over her shoulder.

"Let's walk to the end of the island. There's so many stars I think we can leave the lamp right here. Pick it up on the way back."

"Sometimes," Alice said, as they set out, "sometimes I really am happy, in spite of the ache, in spite of everything."

"That's doing it," Raven said, massaging her shoulder through the thick sweater. "That's the idea."

"Dig it," Alice said, laughing.

NOVEMBER

I drive them down to the ferry. I can't help saying "give my love to dad." It makes them uncomfortable, I know, and

I'm sure the message is never passed on. Then I drive back up the island. Two whole days to do with as I please, except for the dog, except for the cat, except for the hens, and keeping the wood stoves going.

I've been dozing and listening to the opera. Today it was *Macbeth*. What a variety of women Shakespeare created. Lady M. is truly awesome. Is her childlessness supposed to be symbolic? She certainly treats her husband like a child, or a fool. But she gets hers.

> "It is the cry of women,
> my good lord."

"The gang" was over last night. Trudl and Glenn now very much "in love." They both play guitars and sing Leonard Cohen songs and Beatles songs in voices so soft you can hardly hear. Glenn is only nineteen and he has already lived with one woman, is now living with another. Harold got fed up with all the guitar plucking and started stomping around the house opening and shutting cupboard doors, looking for something to eat. Finally he talked Anne into making oatmeal cookies. Lots of clatter and banging in the kitchen; lots of swaddled singing in the front room. Stella sat on the floor by the stove, helping herself to tobacco from the Player's tin, not saying much. What really goes on in her head? She lives up there in that cabin, that *shack*, with a deaf man whom she is obviously fond of but doesn't love and that enormous Saint Bernard puppy who is a bit like Harold himself—large, all over the place, sloppy. She looked restless and bored. Brought me *Love's Body* to read. Selene left me *Stranger in a Strange Land* and Peter recently left *Be Here Now* and *Magister Ludi*. (By their books shall ye know them.) Hannah helped Flora and Christobel build with Lego. She is very good with children and has a kind of detached, yet warm, patience which I don't have. Perhaps because they are not her children?

I miss Raven, not just his help or his hugs but maybe his innocence? In that he is like Harold, a quieter version — he accepts what comes. Harold is stronger because he doesn't need some guru with an eastern name to teach him about acceptance. Harold really loves life — his red hair is just a wonderful accident that fits him nicely into a long line of life-affirming heroes. Stella watches him, with a little smile on her face, starts talking to me about California, how she must go back soon. Glenn and Trudl stare into one another's eyes — "Like a bird/on a wire — "

I've been thinking of sawing my bed in half, then maybe I wouldn't notice how big it is. No matter how I sprawl I just can't fill it up.

Alice received a letter from Raven about three weeks after he had gone back to Vancouver Island.

friends, friendly people seem to be attracted by love, which shows for one another in weird and wonderful ways. All that happens must be accepted as it is, past karmas must be worked out, and in our state (space) who can possibly say what went on in our past lives (blahblahblah) i love you so very very much. just read *female eunuch,* gained a bit more insight. it's only natural for you to love someone as lovable as peter, seems most people do. don't freak though. keep on loving (not grasping) what is offered you from his heart. learn to love it all. the use of living is love and much more of course god truly knows, cause he's all truthfull, be truefull first to yourself, then the rest will fall into place. it's really quite simple if we'd let it be simple. but we are in a strong current of opposition, so at times we can get awful complicated. pickin salal is chilly fingers these days. WORDS!! Comin south by water soon keep the home fires burnin Alice, me here, i love you. were on the side of the lord who will care for us.

Namaste

RAVEN xoxoxoxox

NOVEMBER

"It was the strangest thing," I said to Anne-Marie, all those long months ago. "Maybe it was precisely *because* I was so happy when I went in that I felt it. It just hit me. At first I felt as though the house had been robbed. But it wasn't that. I sat at the big table and wept because I *knew*, suddenly and overwhelmingly what it was. Peter's 'presence' was missing. Not because he was over on the island but because he had 'removed himself' in some way. I realized that he was probably having an affair with somebody."

Anne-Marie bit off her thread. She was finishing up an appliqué for her summer-school class.

"An affair?" She threaded her needle with a new color. I nodded miserably.

"Well I hope it's just that. I feel maybe he's in love, really in love, with someone else. You know, his lovemaking has been so different these past few months, so much freer and more sensual. I thought it was me, or maybe even all the dope he's been smoking. Now I think it's really that he's pretending I'm somebody else. It isn't *me* he's making love to at all!" I began to cry.

"What if he is in love with someone else? *You've* had the odd affair, you know you have."

"I've never been in love with anyone but Peter, never. Oh once, years ago, I fell a little in love with a fellow student. I went to bed with him and he was impotent and we both cried a little. The next week he got married to his girl friend and a little later on he died in a car crash. No one ever knew.

"No," I went on, "what scares me is this sense that whatever Peter's involved in, it's very serious."

"What are you going to do?" Beautiful neat little stitches. In out in out in.

"I don't know what to do. I love him, Anne-Marie. Today I was realizing just how much. I was thinking of all the lovely things he's made me over the years and really, what a good life we've had together and now even the sex coming right. Do you remember that story we read together last year for your English course? 'Bliss'?"

"Yes I remember." She went on with her sewing all the time that we were talking. A little basket of colored silks lay next to her teacup.

"Well, what's her name, the wife, suddenly realizes she's in love with her husband, sexually, really desires him. She can't wait for everybody to leave so she can go to bed with him. Then she sees him, in the hall, making a rendezvous with her new friend. Well that's how *I* feel—so turned on and yet at the same time I have this sick, empty feeling that it's all over between Peter and me."

"I really can't believe that."

I smiled. "I can't either, but I'm so afraid it's true. So many remarks come back to me now, to haunt me. And a postcard I saw in his wallet, once, by accident, when I needed money to pay the paperboy."

"Are you going to confront Peter?"

"I'm afraid to. I'm afraid of what he might say. I guess I'll just go back and hope it will all work out. Maybe he *needs* an affair, maybe that's all it is. But he's going to be alone here in town four days out of seven. Alone and maybe lonely. I can't help wondering if he hasn't 'arranged' for me and the children to live on the island."

"Oh come on, Alice, that was *your* idea."

"I know. I'm just being paranoid. But *I* should've said, okay, then we'll all wait a year. I may have been very stupid as well as very selfish — putting my wants and needs first. I get so impatient. Maybe what I should do is go back and say that now. That we'll all wait another year. The girls will be upset but this is important, whatever it is."

Anne-Marie walked me partway home in the summer night. We arranged to have a night out together when I came in to finish my novel.

"I can't wait to get the damn thing done," I said. "When I have to leave something as big as that it's like trying to interrupt a pregnancy and then take it up again three months later. I'm always scared the little creature will have died."

"I don't know how you do anything at all. I think you're amazing."

"I am, I am."

We parted under a streetlight and I walked home in the soft summer midnight, comforted. Because I knew, of course, somewhere way inside, that she was the one. I knew that and I felt that now she would have to call it off. Now that she understood. Now that she knew how much I cared. I kept it all pushed very far down. My knowing.

And if I had, in the end, said nothing? Had not challenged him? Where would we all be now?

There was an eagle and his mate, Mr. and Mrs. Eagle, high up on the ridge in back of them.

"Did you know that eagles mate for life?" she said. "No, I didn't know that," he said in his new, soft soft tolerant hippie voice. Or maybe he said, "Is that right?" his mind on other things. Seeing what she was trying to do with her sad analogue, but not wanting to be a part of it.

Later on, there was the bumblebee.

Over in the next property, lying in the tall grass beyond the old refrigerator, where he and the storekeeper smoked salmon and cod. Watching himself go in and out of her, smiling.

"You have to understand," he said later, "that what Anne-Marie and I have is [pause] different. Higher."

That night, after she drove him to the ferry, smelling her underpants and crying.

NOVEMBER

If you'll be M - I - N - E mine
I'll be T - H - I - N - E thine
And I'll L - O - V - E love you
All the T - I - M - E time
You are the B - E - S - T best
Of all the R - E - S - T rest
And I'll L - O - V - E love you
All the T - I - M - E time

(Rack 'em up, stack 'em up)
Any old TY - IM
shave and a haircut
two bits)

So little Alice, aged eleven, at Camp Amahami, marching along with her bedroll on her back.

And so big Alice, heading up the island with her daughters. Is it right to teach them such rubbish? To sing about a man without a woman (is like a kite without a tail; is like a boat without a rudder; is like a ship without a sail—) and conclude that if there's one thing worse in this universe it's a woman.

I said a woman
I mean a woman without a ma-an.

Aren't I trying to demonstrate that ain't (necessarily) so? Oh well, it was only a song. It is the singing that was important, not the song.

''Listen,'' said Alice, ''here's a silly one.''

She waded in the water and she got her toes all wet.
She waded in the water and she got her toes all wet
She waded in the water and she got her toes all wet
But she didn't get her [clap/clap]
WET
YET.

''Oh look,'' said Hannah, ''there's a deer.''

''Oh. And another one.''

Two bounds and the deer were so far into the trees they could no longer be seen.

The day Raven and the children made candles Alice had gone by herself to the south end. Perhaps it was the day the booze boat came in and it had been delayed. Or why would she have been away for such a long time? There was no liquor store on the island but it was possible to phone over to Mayne Island, which had a liquor store, and have your alcohol delivered once a week by water taxi. Perhaps that's where she had gone and the car trip was just an excuse. Perhaps she drove down the island, turning off just before ''the village'' (a post office, a gas station, the lodge, the real-estate office, the ferry dock and a general store) and from there to Bellhouse Park. Got out, sat on a bench and watched the ferries going in and out of the pass. With Raven there the kids wouldn't worry if she were away for a considerable length of time.

Whatever the reason, it was dark when she drove up to the bottom of the path. It was nearly Christmas and she was wondering what to do, how to get through it? Peter too — how was he going to manage? He liked Christmas even more than she did, loved parcels arriving from England and the United States, bought ridiculous presents and games, ransacked cupboards looking for the Christmas cake which Alice had hidden. Peter had been a child during the war — Christmases were spartan affairs. And still his parents sent him shirts and ties and underwear. Good British quality from Marks and Spencers.

When he was five his mother had managed to get a chocolate bar for his Christmas. He found it and ate it and she wept. It was one of her favorite stories. Should Alice invite him—would he want to come? They had had wonderful house parties over here at Christmastime. Turkeys so big the door of the cook stove had to be wedged shut with a piece of wood. Bread sauce. Ham. Charades and laughter. Could she manage all that? And if Raven were here would he try and talk them out of turkey? Would they have a "feast" of yams and brussels sprouts? Perhaps they could just ignore Christmas this year — or Peter and the kids could do something in town.

She started up the path and as she got near the house she smiled. There was a light in every window, a candle, and she could see the gray smoke from the chimney rising straight up in the starry sky. The little house itself was in darkness but as she opened the door everyone shouted surprise! surprise! and Raven lit the lamps.

"How beautiful," Alice said. Her eyes were full of tears.

They had been making candles all afternoon, dipped ones and rolled ones that looked like old-fashioned door bells. "Raven said those were your breasts" (and everybody laughed). "We nearly set the house on fire with one of them," he said. "The wick fell out of the candle and 'plop' it went right onto the stove."

"You should have seen the flames!" Everybody laughed

again so Alice laughed too and pushed back a sudden vision of the burning breast igniting the whole cottage, her children with their hair on fire, screaming.

(Ladybird, ladybird fly away home. Your house is on fire and your children —)

She sat down suddenly. Raven took her hands in his.

"It was all *right,*" he said.

"She's such a worrywart," Anne said.

"I wonder why a worry*wart,*" Alice said and they all smiled to see she was her old self again.

That evening, after supper was over, they blew out the lamps and just moved about by candlelight.

"The wise virgins," Alice said, then had to explain.

Raven meditated for a while, the highly tinted photo of his guru, whom the girls insisted on calling "Purple Thing" propped up against the windows in the front room. Raven was having some difficulty with Purple Thing, for to be a true follower he was going to have to give up dope. So far it was a losing battle. Hannah, Anne and Flora made shadow pictures on the wall. Then they all put on their coats and shoes and went for a walk. Alice wanted them to come up the path as she had done, to see their small house shining in the darkness.

They put Flora to bed and sat by the fireplace in the original sitting room, an old army blanket stuffed against the bottom of the front door, to keep out some of the drafts.

"I have a little girl about the same age as Flora," Raven said.

"You *what?*"

"A little girl about her age." They had been playing Rook with a lot of concentration but now the game stopped dead.

"Where is she now?"

"I don't know. She and her mother are in the interior somewhere, the Slocan Valley, I think."

"You *think,*" Anne said. "How come you don't *know?*"

"I tried to find her once, from their last address. My

mother used to hear from time to time. But it was like they had vanished. Nobody knew — or nobody would tell.''

''Why wouldn't they tell?''

''Oh, I was pretty crazy, man, when they left. Pretty weird.''

''What was her name? The little girl.''

''Veronica. Nikki. She'd be a little older than Flora, about five.'' Alice and her daughters stared at him. He grinned, showing all his missing teeth.

''I was doing too many drugs. Hallucinating. Seeing little purple men and all that stuff. Screaming my head off. I wasn't very nice to be around.''

''Was she pretty, your wife?''

''Very pretty.'' He smiled again. ''She had that black hair that's really blue.''

''Then she's the one who should have been called Raven.''

''Maybe.''

Raven began to roll a joint. ''Someday we'll probably bump into one another. If it's meant to happen.''

''If it's your karma,'' Alice said, and was immediately sorry she'd said it. Who was she to make fun of this man? What right had she to judge him? He had just spent hours playing with all three of the girls, accepting them and loving them. It had been his idea to put a candle in every window, they said, and it had been, for her, struggling in the icy waters of Peter's rejection, like a message, all that light shining forth, letting her know something important, something she was likely to forget. Alice smiled at him in apology.

Anne offered to roll out his bedroll in the front room but he said, very quietly, ''I think I'll sleep in the big bed with your mum tonight.''

What a noisy house! The clock on the mantel trotted out the seconds with iron hooves. The wind grasped the old windows and shook them until they rattled. The fire was a whole forest of burning antlers. The cat roared. And all the while Raven sat there, cross-legged, smoking a joint, stroking the cat with his good hand. And what an enormous hand it was, too. Huge. It had more than the allotted number of fingers, not less. Avoiding her daughters' eyes she went into the kitchen and held onto the kitchen table. His hands followed her in. A breast-shaped candle was burning in a blue saucer.

What were they thinking. What would they think? Yesterday he had said to Anne, "I'd like to suck your toes." She had told him to get lost. Was that what Alice was supposed to do? Tell him to get lost? She blew out the candles in the kitchen, licking her fingers and pinching the wicks, then crossed the room where Raven and the older girls still sat, went into the new room, blew out the candles there, put on her flannel nightgown and got into bed. She called good night but there was no answer.

Soon she could see the shadows of her daughters against the curtain, hear them in the bathroom washing faces, brushing teeth. Raven had gone out to bring in wood for the morning. The girls blew their candle out so now there were just the patterns of firelight from the chinks in the pot belly stove.

"Good night," Alice called again, determined. They mumbled good night and then there was only silence and the sound of the wind outside. The whole house waited. Even the mice in the cupboards were quiet. "Twas the night before Christmas," Alice thought, "and all through the house. . . ."

When Raven came in he lit the candle by the bed.

"Don't," Alice whispered, "please."

"Yes," he said, "I want to see you." He put some more wood in the stove, then took off his clothes and came to bed.

It had been weeks since any man had touched her with affection. She realized that some time during the afternoon Raven had taken a bath and washed his hair. His skin was a lovely color in the candlelight.

"Your hair smells of smoke," he said.

Alice laughed. "So does yours. Of smoke and dope."

He pulled her nightgown over her head and then lay down again with his face against her breasts.

''Your breasts are lovelier than the candles,'' he said, and began to kiss them.

It was only when he turned his head away a little, toward the light, that she could see he had been crying.

Fruits de mer, fruitti di mare, Alice's mother fussed and fretted about how they were going to manage, out there on that island, with the husband gone.

She traded apples from her apple trees for fish — cod of all kinds, red snapper. And Hannah and Anne jigged from the rocks with a hand line. The rowboat seemed somehow to be Peter's boat exclusively and so they left it alone. Or until the Easter vacation when they took turns scraping it and painted it Chinese red. When the tide was out they dug for clams, working quickly before the clams dove downward in their streamlined shells. Alice put them on the back porch in a bucket of water sprinkled with cornmeal until their stomachs were all cleaned out, then made up enormous pots of chowder. They pulled mussels off the rocks, cooked them in their own liquid with garlic and butter. Collected oysters. Mushrooms (only a few of these, only the ones they could be absolutely sure of). Turned over the garden, ready for spring, picked out the stones and piled them in one corner. Sprouted seeds in jars, made yogurt.

And when Peter came out or the girls came back from town there was always something sweet and sinful to make up for all this health. A Black Forest cake, for instance, oozing cream and chocolate. The kids called these treats ''gooey yummies'' and although Alice enjoyed them, licked her fingers with the rest, she found them rather peculiar. All that oozy sweetness, cakes covered with liqueur cherries or sugar roses. There was something sick and sentimental about them, something not quite right.

With a borrowed juicer, when Raven was visiting, they made apple juice and cider from some of the apples. Alice thought she'd like to learn how to make beer.

Another friend had lent them a Salish spinner and before the breakup, Anne-Marie had sold them a bag of wool. Hannah was the only one who could make the spinner work properly so the rest of them teased and carded the wool, sitting on the rug in the front room. Selene had given them a small jar of lanolin before she left for New York so they took turns giving each other back rubs. Just playing around with the wool made their hands soft and smooth.

They made a puppet-theater - television-set out of a large cardboard box and gave shows. Stella and Harold, Trudl and Christobel and Glenn walked down from their cabins and everybody dressed up. Sometimes Alice could not believe how happy she was. Sometimes she felt guilty for being so happy. How could she be so happy when her heart was broken? And it was; there was no doubt about that. Sometimes, waking in the middle of the night (the foghorn out there with its terrible lament, something so sad you couldn't put words to it, you could only moan) she could feel the sharp broken edges of her heart grating against one another so that it hurt to breathe. The nights were long that winter and they went to bed early, partly to save wood, partly because (or so it seemed to Alice, looking back later) they each needed some time to be alone, to not speak, to not have to "relate." Each in their wedge of darkness, thinking what thoughts? While Raven was there he broke up the intensity; he was a new element in their circle. And he was male — there was that, too. He introduced some maleness into the intense femaleness of the situation. Like a traveler from some far country who comes to the door on a stormy night, he sat down and accepted their hospitality and in return told them stories.

He told how in high school he had got bored in typing class and threw the typewriter out the window. He told how he used to wear green-suede shoes with pointy toes. He told about "boosting" a canoe from the army-navy store. He and his friend. Just picked it up and walked out with it. Who would think that anyone would steal a canoe? He and his friends had "boosted" lots of things over the years. It was a game.

But that was not why he had gone to prison — or prison farm, really. He had been arrested for dealing in dope. Somebody told. And so he had gone to prison farm for a while. That was cool. They let him work on kitchen detail. He spent a lot of time chopping heads off chickens. That was how come he only had nine fingers. There was a provincial park near the farm and he used to wander over there, smoke up with various people. So one day he was high and wasn't paying strict attention to what he was doing, just chopping away outside, sitting on the back stoop of the kitchen, grooving on the sunshine, and suddenly it wasn't a chicken head flying through the air, man, it was his finger.

"Don't," Anne said. "I don't want to hear about it." (It was Anne's job to make sure they had lots of cedar kindling. Hannah and Alice chopped up fir and left arbutus or pieces too full of knots for Peter when he came over. All three of them did a lot of chopping. It could happen to any one of them.)

The surgeons hadn't been able to save the finger. (Alice wondered why not, whether, if Raven hadn't been a prisoner —)

Telling them all this in his soft voice, pulling strands of wool through his strong fingers, the brightly tinted photo of Purple Thing propped up against the window behind him.

After he left he sent them all a letter, folded around two joints. (Alice laughed; she hardly ever smoked dope. She would give them to Stella and Harold or leave them for Peter. Thought of Marx's dictum, "from each according to his abilities, to each according to his needs.")

This was Raven's letter written in Magic Marker:

> Do you have a pain?
> God will accept it if you offer it to Him with Love.
> It really works with anything from a cut finger
> to deep heartache.
> Just give it up.

And then there was something that must have come from a book.

> The fewer the Words
> Less the Artifacts
> Slowly the luck of it
> The knack for non-attachment.

That was all very well, thought Alice. He and Selene would soon be back together. The knack for nonattachment was much easier to acquire if one were attached, quite firmly, to someone else.

NOVEMBER

He gave me a kiss at the door. He and the children were going back over to the island so that I could finish the last bit of my novel. (I had taped one of my inevitable lists to the refrigerator door at the cabin, happy to have a few days by myself. "5. Love one another!") He said he wanted to have intense relationships with other people.

"Peter," I said, "do you mean intense relationships with other women?"

He gave me another kiss. The children were already in the car. He tapped me on the nose.

"Get on with your book."

Two days later he and Anne-Marie were weeping in each other's arms, deep in the woods, on the crown land that he and I had staked and hoped to claim. Weeping desperately like the woodcutter's children, wondering if there was any way out at all.

Our petition for the crown land was denied and I was just as glad. I have never been back there, to that part of the woods. I do not want to walk around thinking, "Was it here? was it here? was it here?"

MENE, MENE, TEKEL, UPHARSIN

(You are weighed in the balance and found wanting)

Alice found that message in a drawer. She could not remember where she had copied it from, or from what language.

Alice and the children were playing word games:

"Una Lateral."
"Una Nimity."
"Una Versal."
"Una Verse. A poet."
"Una's son."
"That's a good one."
"Una T."
"Una Form. A Sexy Lady."
"Una Corn. That's you mother."
"Very funny. I can't think of any more."
"Let's play Categories instead."
"Let's switch to another prefix."
"What?"
Alice thought, "Auntie"
"I don't get it."
"It can mean before or against depending on the spelling. For example, 'Anti-Christ' becomes 'Auntie Christ.'"
"That's too hard. You know too many words."
"Let's play Categories instead."
"'Catagory'—now there's a good word. Somebody ran over the cat."

"Mother."

"All right. All right. Have it your own way."

And then they all had hot chocolate with marshmallows floating in it and went to bed.

Alice picked up her sleeping youngest whose cheeks were flushed from the fire.

"I *like* them," she thought, "I not only love them, I *like* them. That's the miracle." Then she put the cat out and let Byron in and padded off to bed mother-catlike, in her soft green dressing gown. She wondered what Peter was thinking about right then. Did he miss all that or had he been fed up with it? She tucked her feet against the towel-wrapped hot-water bottle. She really felt like getting up again and phoning Peter, just to tell him how good she felt tonight. She knew he worried about her. She even knew that in some very deep, essential part of himself he loved her still. One of her friends had written, of the awful things Peter had said to her, "That's the exaggeration of desperation." She had to hang on to that.

It was snowing again, in big flakes. She lay on her right side so that she could fall asleep facing the windows. White crumbs falling falling falling. The woodcutter's children. Joyce's story "The Dead."

Go to sleep Alice, she told herself, turn it off. Whirr. Whirr. Whirr.

The dead are so terribly dead, the lonely so terribly alone. Alice, playing with words even as she fell asleep. Lone-lie. Only one. And thickening, thickening, no longer the sapling-like creature in the Bell Hotel in Dublin, writhing in ecstasy on the landlady's lavender-scented sheets.

Years later, on the island, remembering that place, Alice planted her herb garden underneath the clothesline. Sage. Rosemary. Mint. Lavender. Borage. ("I borage bring you corage.") Everything smelled lovely. But who was there to smell? Where was the other figure from that bed in the Bell Hotel? And what was the smell of a lavender pillow compared to the lovely tropical scent of come and cunt mixed up and running down her legs?

She read about a flower called love-lies-bleeding.

"About this weekend," Peter said on the phone. "Are you still planning to come in here?"

"Of course, I need a break. Why?"

"Calm down. I'll be over on Friday night." He paused. "It's just that Raven and two of his friends are staying here. They're nice but they're pretty weird. The girls upstairs said they would be away and you could use their place."

"I don't want to use their place, I want to use my place. I thought you moved out because you felt I should be free to come and go and have some peace."

"I did, but this was an emergency. Actually, if you wanted to switch weekends and come the following one, they'll be gone." He laughed. "They're leaving on Sunday. They're on their way to California to pick avocadoes and follow Kirpal Singh."

"That's nice," Alice said, "but I need a break *this* weekend. I *need* to be alone."

"Well, I don't feel I can tell them to go."

"You shouldn't have told them they could stay."

"I really didn't think you'd mind."

"Oh come now, Peter, you know what a bitch I am, you knew perfectly well I'd mind. Now you're testing me once again, to see if I can move with the changes or whatever nifty phrase is in vogue this week."

She could hear his sigh. "I'll tell them to go."

"No. No. Of course you can't. Although Raven would understand, I think. I'll just have to put up with it for now but I'm not your wife any more and any decisions about house, cabin, *anything* like that had better be joint. I don't know about Raven, sometimes he scares me. You know he admires Charles Manson, don't you?"

Peter sighed. "He doesn't admire him. He says he understands him."

''No he doesn't. You are so far into your romantic shitty dream you don't listen; he said that it was those people's karma to die and Manson was just an agent. 'It was their karma,' that's what he said.''

''Well, he may say things like that but he's really very gentle.''

''Oh just fuck right off Peter. I'll be in on Friday night. They can stay until Sunday and then *out*. You want to be Mr. Good Guy don't you? I notice they're not staying with *you*. But I won't spoil your image.''

She slammed down the phone.

Alice had been invited to have dinner on the Friday with some artist friends. When she arrived at her own house the record player was turned up loud and there was a strong smell of incense and dope. A candle had been stuck onto a low table Peter had made and the wax had made a large frozen puddle on the wood.

Nobody seemed to be around. In the kitchen was a bag of apples, some yogurt and a saucepan full of cold brown rice. The sink was full of dirty dishes.

She made herself a cup of coffee and wondered if she should go over and stay with her friends for the night. Was sitting eyeing the telephone, when a young man in bare feet and blue jeans walked in.

''Howdy,'' he said in the usual soft voice. ''You must be Alice.''

''That's right.''

He plugged the kettle back in without offering *his* name.

''You really shouldn't drink coffee,'' he said, ''it speeds you up.'' He helped himself to some Red Zinger tea which Alice had bought the last time she was in town.

''Really,'' she said. She tried to say it the way they did, as a pleasant confirmation, ''*Really.*'' Unfortunately, it did not come out that way.

''I also drink Red Zinger,'' she said. ''As you know.''

Her hands were shaking and she had to lean over and slurp her coffee into her mouth like a cat.

"What's your name?" she said. "Sunshine? Or Heron? Noah?"

"Henry," he said.

"You really ought to change it."

A girl came in; she was Henry's about-to-be-wife, from the south. "Mae Love." She helped herself to some tea and sat down at the end of the table. She and Henry began a discussion about some big cosmic meeting that was coming up.

"If you're not listening to the record player," Alice suggested, "why don't you turn it off?"

"Paul Horn in the Taj Mahal."

"I know that. It's my record . . . our record," she added. Why am I like this? she thought miserably. Why should I let him bother me?

"Raven's listening," Henry said.

"Raven? I was just in the sitting room and I didn't see him."

"He's there."

"Meditation," Mae Love said. "Honey, is there any of that apple butter left?"

Alice got up and went into the sitting room. Raven sat cross-legged on the couch which was hidden by the French doors. His eyes were closed.

She went back to the kitchen, said, "Excuse me," and reached for the telephone. While it was ringing she said to Henry, "Why don't you do up all those dirty dishes while you're waiting for Raven to finish?"

"The soap's all finished," Mae Love said.

"Buy some," Alice suggested. "I'm sure they sell organic soap wherever it is you shop. Or make some. You do it with wood ash. I'll show you how."

Mae Love looked at Henry. He raised his shoulders and shrugged. They got up and moved out of the kitchen.

Alice remembered Selene arriving at her cabin.

"They don't care," she said. "There was cat shit everywhere. I'm the only one who ever cleans up. I have to leave."

"You get those dishes cleaned up before I come back," Alice yelled. "Or out you go!"

Nobody answered her phone call. She picked up her cape and purse and walked out. When she slammed the door it made the record jump. She wondered if Raven's inner eye flew open at the sound.

Out walking, trying to simmer down about Raven and his friends, Alice saw warm glows behind the drawn curtains of ordinary-looking homes. She knew that's where she really belonged, cutting recipes out of the evening paper, kiddies asleep or downstairs watching television in the family room. Tears stung at her eyelids because no, she didn't fit in *there* either. Not without some kind of lobotomy. But where? Was everybody just playing games? Was anything really real? She remembered the Cheshire Cat—"we're all mad, or we wouldn't be here." *That* Alice woke up and it was all a dream. She felt alien and afraid and pulled her wool cape closer to her. Walking the quiet streets, the yellow lights of the houses as innocent as butter pats. Dragging her steps a little, wishing some Great Parent would swoop down and gather her up with a cry of "Oh *there* you are" and tuck her into bed and kiss her cheek and tell her the world would soon be right again.

She walked for an hour then let herself in and went into the bedroom. Raven was asleep in her bed. She wanted to make a joke of it. She wanted to say, in a deep, gruff voice, "Who's been sleeping in *my* bed." Instead she began to cry.

Without opening his eyes he said, "Take your clothes off Alice and let me hold you."

She wept in his arms for hours while he said nothing, just stroked her hair.

And then they began to make love. "Oh Raven," she wept, "I'm so *lonely*."

"Hush," he whispered. "Hush little sister, no words now. Let me love you." (A second Alice watched them laughing cynically.)

On the Sunday morning, backpacks ready, Raven and

Henry and Mae Love and Alice sat in the sitting room. They were having one last cup of Red Zinger. Alice had lent Raven some money so he could get across the border. He was sitting in an armchair and she was on the floor, her head against his knee. His gentle fingers were massaging the back of her neck.

"'One needs a teller at a time like this,'" Alice said.

"What's that?"

"It's a line from a poem. First world war, I think. No, maybe second. He meant that we all wanted some power to tell us what to do."

"I can dig it," Raven said.

"You know I think you're bananas, don't you Raven?"

"Ummm."

She wanted to tell him not to go.

"Let's boogie," Henry said. Raven got up.

"Peace, my sister," he said.

"I hope you find what you're looking for."

"I will. I know that."

She stood at the window and watched the three of them set off down the street.

Then turned on the record player and curled up on the couch which Raven had used for his meditation.

"We're all mad," she thought, and shut her eyes, "or we wouldn't be here."

NOVEMBER

"'It just comes in the wintertime,'" the old Indian woman said. "You can't sing in the summer. It don't like it here and went back to his home. He stays with you all the time but in the summertime the song goes away, it goes back to where he stay all the time like he was a person."

Then she added:

"Sometimes on his way the song gets lost."

Your soul can be taken by the ghosts and then the doctors have to go on a trip to get it back. (Raven would like that.)

"One uncle he said he was down on the beach and he called 'herrings, herring, herrings.' All the ghosts came down with their herring rakes and tried to rake herring. While they were down on the beach the other uncles went in and got the the soul of the woman and dragged it back with them."

"Once at a power dance X put what was supposed to be P.G's daughter's soul back on her but it burned her head. After they got home P.G. took it off. It was really a ghost."

And then we have Emily Dickinson, "One need not be a chamber to be haunted — "

Sometimes, very faint, I hear the drums on Kuyper Island, the dancers practicing. I want to go to them; I want to say, "Help me to dance away my grief."

They spread Indian cloths on the rug and everyone sat on the floor. Alice, at one end, her back against the bed, looked down the length of the room and tried not to think of Mrs. Ramsay, staring down her dinner table, asking herself, "But what have I done with my life?"

"Is daddy coming for Christmas?" Anne had said.

"Shall I ask him?" Hoping they'd say yes. They wrote him a funny invitation. To Mr. Ebeneezer Hoyle. He said he'd pay for the turkey and bring it out and all the veggies on Christmas Eve. On Boxing Day he and Anne-Marie and maybe Jeannie planned to go with the older girls to Hollyburn Lodge for a brief holiday. That meant they'd have to leave the island Christmas night.

"Oh that doesn't matter," she said, "we all want you to come. We can't imagine Christmas Day without you." She was playing on his guilt but didn't care. "We're making a piñata.'' (Charming way to smash something if smashing proved to be necessary.)

''Very well.''

Just to be on the safe side Alice invited two other couples they knew. It would be a house party, like old times, everybody sleeping wherever they could. Stella and Harold were having Christmas dinner at her father's but said they'd come over for the piñata breaking. Trudl and Glenn and Christobel were having dinner at Stella's father's, then catching the ferry into town to visit Trudl's mother and leave Christobel for a while.

Trudl was crocheting beautiful woolen caps for everyone, all in rainbow colors.

''We'll look like a team,'' Anne said.

''We are.''

Alice made old-fashioned flannel nightgowns for the girls. She wrapped up *The Joy of Sex* for Trudl and for Stella and Harold Chinese rice-patterned bowls and fancy chopsticks. For Stella alone *One Hundred Years of Solitude*. She was making play dough for all the little kids but didn't know what to give Peter. Finally she decided on a new fishing pole and went into town to puzzle over it at the army-navy.

''If Raven had been with you he could have boosted it for you,'' Anne said. Hannah made shortbreads, Anne made potpourri from the summer's roses, Flora made God's-eyes out of crossed sticks and scraps of wool.

What else could Alice give Peter? What else? She went back into town on her last free weekend and bought nuts and a nutcracker for Christmas Eve, Diane Arbus's *Freaks* and a box of chocolate-covered ginger. Paper water flowers for the stockings, sugar mice, balsa airplanes, little puzzles, new toothbrushes, an entire farm in a matchbox, a set of miniature playing cards, windup elephants and turtles. Peter's mother had always done the stockings. This year she had written and said she couldn't but had sent a parcel and to watch for it. They were all tremendously excited.

''Who's going to get the tree? Daddy always does it.''

"Well for God's sake we can chop down a little Christmas tree."

"Let's have a live one."

"I'm glad there's no vegetarians coming. I'm *glad* we're having turkey."

"What are you going to eat Hannah?"

"I'm just going to eat the salmon."

"I'll bet you'll eat the turkey too."

"Can we dress up?"

Harold and Stella found an old wicker-seated sled in the Goodwill and fixed it up and gave it to Christobel and Flora.

Everybody prayed for snow and by God, it snowed.

Alice, on her knees with a scrub brush, scrubbing the gold carpet, looking at all the wine stains, candle wax, burned places from sparks. It had had its initiation, that was sure. They decorated the tree, a live one set into the big preserving pan: paper chains, cranberry strings, popcorn, gilded walnuts and gingerbread birds, elephants and camels, cut carefully around cardboard patterns Alice had made her first Christmas in Canada.

"Aren't we going to have any lights?"

"What do you think?"

"I wish we could have candles. It must have been lovely when they had candles on the trees."

"I wonder how many caught fire?"

Glenn and Harold chopped extra wood for them in the morning of the day before Christmas Eve, and then the team, wearing their rainbow hats, the older members more than a little drunk on mulled wine, took turns pulling Flora and Christobel up and down the snowy ridge road, made angels, threw snowballs and sang carols at the top of their voices. Byron and Uggah peed everywhere in their excitement. The cat followed daintily, complaining.

"We three Kings of Orient are
Puffing on a loaded cigar — "

They made a large anatomically correct snowman in the yard in front of Stella's shack and gave it a rainbow hat out of food coloring. Its genitals were two wrinkled potatoes and a blunt-ended carrot.

"The Indians will be talking about us again," Stella said.

Then Christobel wanted a snow lady in front of her house, Flora wanted a snow dog and cat in front of hers and by the time they finished it was dark. Alice lit the tree and turned off all the lights. They stared in at it from outside.

"It's beautiful."

"There's something magic about small colored lights," Alice said. A second tree was made from the reflection in the windows. It was beginning to snow again. She hoped she wouldn't be to scared driving down tomorrow.

"Oh," said Christobel, when they all trooped inside to have soup and brown bread. (Alice had covered the floor in the entrance room with old newspapers.)

"Oh my gracious." She hadn't seen the piñata before. It was a gigantic flower-decorated pig.

"I love you Christobel," Trudl said, "you're such a funny little old lady."

Christobel buried her face in her mother's neck.

"I love you too."

That night, lying in bed, Alice said over and over and over to herself oh please God please God please God please. The holes in the pot belly stove made pretty patterns on the ceiling.

Alice was in the kitchen trying not to cry. The children had their jackets on and were waiting. Peter came to her and she turned and wept into his collar.

"Hush," he said, "hush there. It was a lovely Christmas. I think the best we ever had."

Trudl was driving them down so Alice wouldn't have to make the trip twice.

"Oh Peter!"

"It's all right." A car honked. "We've got to go."

If the children hadn't been there she probably would have fallen on her knees and begged him not to leave her.

Late in the evening he called. ''I just wanted to thank you again for the lovely Christmas.'' She could not speak at all, just nodded dumbly at the phone, and tried to smile and finally put the receiver (gently) down.

Section II

''At every turn we encountered new obstacles.''

—*A Spanish Voyage to Vancouver*

''Vegetables,'' Alice said to the cat, ''tonight I am going to have vegetables, not veggies. I may even put some chemicals or irritants on them. I am also going down to the store to buy a ridiculously high-priced piece of frozen meat. But I am not about to have a 'feast.' I am going to have my *supper*. And then, in a smoke-free house, I am going to get undressed and read in bed while listening to the radio. You may join me if you wish. If you stay at the foot of the bed and don't scratch. If the phone rings, if there is a knock on the door, we will ignore it.'' She looked seriously at the cat and at Byron, who seemed to be grinning at her.

''That goes for you, too. No barking. Restrain yourself.''

Once in a while there was something to be said for utter solitude, even if she knew her contentment was partly based on the fact that her solitude was temporary, that the girls would be back on Sunday night. Meanwhile, tonight at least, she could be mother to herself. Raven was back on Vancouver Island, Selene was in New York, chanting mantras. Stella and Harold had gone to Victoria. Trudl and Christobel and Glenn would probably not come out tonight, it was raining. Alice liked the sound of the rain on the roof.

She ate her dinner listening to a Benny Goodman concert on FM. Byron was put outside to chew on the steak bone while Alice carried everything that she thought she might need into the new room. A bottle of dry sherry, an apple, some digestive biscuits, the radio, a *New Yorker* and a *Natural History* magazine. She built up the fire in the pot belly stove and let the kitchen fire go out. She got into bed and put Peter's dressing gown around her shoulders. Lately, when she heard herself laughing or realized she was smiling she had wondered how she could do these things when her heart was broken. Sometimes she felt it was a betrayal of her grief. But tonight she just felt grateful. To be content — if only for a few hours. This was the sort of thing that she had imagined she and Peter would do, after he had moved over. Wasn't it enough that *she* was doing it? Selene was right to wonder, ''What does she *want*?''

She picked up an article on the unsluggish habits of the slug and began to read.

The woman, at the usual time after her delivery, shall come into the Church decently appareled and there shall kneel down in some convenient place —

Alice in front of the stove, kneeling before the open door in Peter's cozy dressing gown (the girls in town), putting another and extravagant piece of wood on the fire. Firelight and candlelight and lamplight — the rest of the house in darkness and the pebbled night flung against the window-panes. Her proofs lay forgotten as she floated in her memories, currents hot and cold, lagoons and clashing rocks. ''The cracked servant of the looking glass,'' she thought. ''No, that's not right.'' And always water. If it wasn't rain it was tears or cups of tea or the salty channel. Lately a general aura of damp about her life, a sense of soggy handkerchiefs. She'd have to buck up, take herself in hand, stiffen her upper lip and put her nose to the grindstone. ''Sheer plod makes plough'' and happy women have no histories. A fact. How long was this obsession going on? This Peter nonsense never petering out. He didn't want to be a Real Boy anymore and could she blame him? Being a Real Girl all the time was no big deal. Much more fun to wing it around with the other

Lost Boys and Girls. Hnnh, Hnnh, I'll huff and I'll puff and I'll — Hold the smoke in a minute and look mom, I can fly. Stoned. Such an interesting word. As compared to ''sloshed'' or ''smashed'' or ''pissed,'' for example. A hard word, ''stoned.'' Who had thought it up? From astonish, probably, so it must have been a professor unless a happy accident. And yet those who really went in for dope-smoking, the ''stonees,'' went in for softness, not hardness. They wanted to be ''mellow.'' ''Consider the lilies of the field, how they grow.'' It is dangerous to lean out the window and please don't raise your voice. Stoned. Huxley said that on drugs anyone could have Blake's visions. That was very democratic of him but Alice didn't believe it for one minute. She liked Hotspur's exchange with Owen Glendower:

> *Glend*: ''I can call spirits from the vasty deep.''
> *Hot*: ''Why, so can I, or so can any man;
> But will they come . . .''

But something in the dope had changed Peter, releasing him from the rock and not the other way around. Peter said that when she picked up a pen she was ''naturally stoned'' and she found the remark insulting. Perhaps it was just the language, the jargon, that bothered her. He said Anne-Marie had remarked that children are ''naturally stoned.'' So what they were all after then was a sense of wonder, of innocence? They wanted to be carefree, to leave the boring adult world for someone else to run. ''We all live in a yellow submarine.'' They wanted to dress up; they wanted to play. There was nothing wrong with all this and yet it made her uneasy and afraid. Afraid for the real children, afraid for the world. It would be hard to take a stand on anything if you were always sitting with your back against a log, your eyes closed, just ''being.''

''At the start,'' Alice read in the book of rules, ''the Queen is the Piece placed nearest the King.''

JANUARY

Mother (1), a female parent. (E.) M.E. *mōder*. A.S. *mōder, modor,* a mother; the change from *d* to *th* is late, after A.D. 1400 + Du. *moeder,* Icel. *mosir,* Dan. Swed. *moder,* G. *mutter;* Irish and Gael., *mathair;* Russ. *mate,* Lithuan. *motė́,* L. *mater,* Gk. *uńtnp,* Pers. *mādar,* Skt. *mātā, mātr-*. Orig. sense uncertain.

mother (2), hysterical passion. (E.) In *King Lear,* ii. 4.56. Spelt *moder* in Palsgrave, and the same word as the above. + Du. *moeder,* a mother, womb, hysterical passion' cf. G. *mutter-beschwerung,* mother-fit, hysterical passion.

Mother (3), lees, mouldiness. (E) A peculiar use of Mother (1).

Mŭmḿy̆ c. 1. Body of human being or animal embalmed for burial; dried up body. 2. Pulpy substance or mass, esp. *beat* (thing) *to a* ~ .3. Rich brown pigment. (f. F. *momie* f. med. L. *mumia* f. Arab. *mumiya* (mum, wax.))

mummy̆[2] , n. Mother (nursery form of MAMMA)

Who can see the "other" in mother? Calling the school for so many years — "Hello, this is Hannah's mummy." "This is Anne's mummy"—to make identification easier for the teacher. All wrapped up in her family.

"Don't you know what a *Bombardier* is?" one of the children said. They had all come down off the mountain in a

Bombardier. ''Anne-Marie was scared,'' they said. ''She kept her eyes shut the whole way down.''

Languid, that was a word that Alice had always particularly liked. Cane chairs and long white dresses. Tall cool drinks. Roses in a silver bowl. Languid conversation. Languid afternoons. Languid poetry. ''Why so pale and wan, fond lover? Prithee, why so pale?''

Now languid — Peter tasting the cool juice from Anne-Marie's cunt on his exploring tongue? Alice obseeing and then obseen. An Etym dub.

And those little biscuits called ''Les Langues du Chats.''

''LOVER,'' she wrote. Then drew a line across the L.

Anne-Marie (in a letter to Alice)

''It was meant to be.'' And then, ''I love you. You are inside me every moment.''

Stella (in the Bengal Room of the Empress Hotel, amidst leopard skins and scimitars hanging from the walls)

''It was bound to happen.''

dear friend Alice —

It is the end of an incredibly lovely day. Incredibly because I am still attached to Time and Space enough to expect this kind of physical environment to have an unlovely effect on me. I worked today from eleven-thirty to three-thirty as a waitress at Schraffts (remember?) and was able to stash away a few more dollars in my ''going home'' stash. I actually like working—it's very demanding physically, my left arm which *always* carries the tray is quite strained and sore and may become quite deformed from such unnatural muscle development but there are compensations. I enjoy the necessity of having my mind perfectly organized for a few hours a day —the rest of the time it can float where it wants—there's no "homework" with this kind of job. Also the tips are good—food is free—although the high sugar-carbohydrate content is doing my body in. Still feel healthy though, so I can't complain. People are fascinating. Always ready for a fight or to be neglected or treated rudely so when I refuse the challenge and instead insist on loving them (in some small way) they become quite passive. It's fun—for a while.

So then I met mother at her office and we went to her health club for a sauna-swim-shower-sunlamp orgy after which I felt purged and clean and ready for an evening devoted to the spirit. Every Wednesday and Friday night, unless I have to work—I go to the home of some very high people for chanting and meditation and news of Baba Muktananda and his Indian ashram. Learning many gorgeous chants which have the power of releasing a person from mental "control" without fear. Many of them bring me to tears with the delicate beauty of their melodies. Tamboora and Indian bell-like percussion instruments, subtle incense and the candles, Sanskrit phrases, Indian carpets—all make me feel very much at home somehow. Once a month, on or near full moon, chanting from noon to midnight one chant —OM NAMAH SHIVAYA—holy name of Shiva. This month it's Saturday the seventeenth. A very good way to clear the mind and senses for to start all over again. So this is how I've kept my spirit from shriveling in this crazy crazy town.

And how is Alice? I remember in the first letter you wrote me you described how you were a fountain from which people would drink and then go away, refreshed. And I thought, what more could she possibly want? I envied you, in every detail: your home, children, friends, creative wealth, success in letting *others see* what's in your heart and head, your relationship with Peter — even that! I have never even *known* a man fourteen years and to be still very much in love with him after knowing him that long and that intensely!! How many people can claim that? Remember . . . it's the love that's the important part. Just keep radiating it in all directions and you will be satisfied in every way.

Soon I must come back to all of you. I yearn for silence and trees and watching water rolling in and cuddling big and little bodies around a fire and *kapusta* (yes, even with an irritant or two). I wonder about Trudl and Christobel and Stella and Harold and Byron and Uggah who are so lucky to be country dogs and able to shit and pee in PEACE! (dogs in New York are so sad). I have had many wonderful messages from Raven and I do love him so much. I wonder what it will be like to see him? I can't imagine falling back into hurtful habits . . . we'll just have to see. I think I will leave here the first week of March and arrive the third week in Vancouver — tentatively of course. Then maybe you could go for a holiday? Raven mentioned you had it in mind. Maybe I'll sail directly to the north end on my magical boat and bypass customs (tee-hee) then again that wouldn't be strictly according to Singh, now would it?

May I thank you for lending me your lovely dress? It has been a great comfort. And so have you been.

Yes, drugs are the key. They are certainly *not* what you find when you finally get the door open!!

The words on your OFFICIAL PICTURE ("companionship") are meant for both you and me and anyone who reads them, really.

Much love and many hugs . . . I remain your (underfelt) —

Selene

One day Raven appeared at the cottage door. He had canoed down to see Alice and some of his friends at Coon

Bay. He wanted to know if they could have a bread-baking session the next day. They would buy all the ingredients at the store if they could use her oven.

"How many of you?" Alice said.

"About six. We just feel like doing it and nobody has a real oven."

The girls were away and Alice was feeling blue so she said okay. Before she went to bed she took a bath and washed her hair in honor of the occasion.

And it was that. Six of them all stirring and kneading, making elaborate twists of dough, adding goodies like raisins and chopped apples, nuts, cinnamon and grated cardamom (no sugar of course, but honey). The oven was very small and tricky—things burned on the top and didn't bake on the bottom. Halfway through the loaves would have to be tapped from the pans and turned upside down. It was all a matter of timing.

On the ash box it said "CABIN COOK." The first time Alice saw it she thought it said "Captain Cook" and remembered how that unfortunate had been hacked to bits (and presumably eaten) by those nasty natives. Brave George Vancouver manages to retrieve some of the bits and pieces. A thigh bone, for instance.

Raven and his friends had come to her in canoes and yet they did not come as enemies. Nevertheless, nevertheless, they made her uneasy. Pulled books off shelves, looked at them briefly and put them back. Turned the tape recorder up high. Everybody had a good time. Everybody got stoned. Alice did not like them lying across her bed, but they had all taken their boots off at the door so what could she say? The small house was full of the smell of baking bread and dope. One of the men, who had an English accent, told her he had been at university and suffered a nervous breakdown there. Now he was living at Coon Bay with his wife and working on the roads and going tree planting in the summertime. They were saving up to go to California and do primal therapy. You had to get back to before the time your parents fucked you up. That meant, usually, going right back, maybe even into the womb. Alice had seen him recently on mail days; he was very good-looking. One of the other men was quite experienced in astral projection.

All of them on quests, longing for answers, the Lost Boys. In what way was Raven any different? Did she only think he was because she knew him better? Who was Purple Thing if not a Parent who could speak for God, a priest? And she, too, wouldn't she like to find enlightenment and peace? Wouldn't she like to lay her anger down? Wouldn't she like to have been a pilgrim with a scallop shell on her cap?

But one went on a true quest alone and, except for magical or divine intervention, one fought the terror of the dark wood alone. One didn't bring along three kids, a lame dog and a spiteful cat.

JANUARY

How many of the men who explored these fog-shrouded coasts and uncharted inlets were actually married? Only Vancouver's gunner on *his* ship and he didn't bring his family along. Questers have to Pay Attention or they miss the sacred signs. But nice to set out on a quest together, "hand in hand," like Milton's Adam and Eve. The dark would not seem so dark, the stars so cold. It might have been the best thing that could have happened to them.

So I sat at my kitchen table and listened to their dreams and aspirations and smiled. Smiled too, to think of anyone looking in the windows and seeing one woman with six somewhat unkempt, mostly bearded, men. How's that for a fairy tale?

"What are you smiling at?" Raven said. It was his turn to bake bread.

"I was thinking that if we had one more man we'd have a fairy tale. It's usually three or seven."

"Well, here comes one up the path," Raven said. "It's Harold."

That afternoon was the first time she ever saw Harold distressed and puzzled. Although Raven spoke directly to him, the other men would forget (or did not know) about his deafness. And there was so much talk, so much going on. She saw him standing between kitchen and the front room turning his good ear, with the hearing aid, first toward one side, then the other, an anxious frown on his face.

At around 8:00 P.M., after soup and bread had been eaten, all the men left, taking with them most of the baking and, as Alice was to discover later, the little hash pipe that Peter had given her. It didn't really matter — she hardly ever smoked anything — but it had been a gift.

Raven stayed behind. Someone else would paddle his canoe back to Coon Bay and he would borrow Alice's lantern and walk back through the woods.

He asked if he could take a bath, so Alice lit the oil burner and shut the door of the bathroom to heat it up. While he was in the bathroom she opened the front door to get rid of some of the smoke, then built up the fire in the fireplace and in the front room.

He came out of the bathroom in just his flannel shirt.

Alice had spread a blanket in front of the fire.

"Your bum will get cold," she said. "Maybe you'd better put your pants on?"

"What if we got into bed?"

"I thought you were going back to Coon Bay tonight."

"I can go later. Or tomorrow. It's no big deal."

"I'm shy."

"With me?"

"No, not with you; not any more. About them. About all of them knowing."

"It doesn't matter."

"I guess it doesn't."

She knocked the logs apart and put the screen in front of the fire.

Cat out. Dog in. Wind up the clock. Lights switched off. A candle. Raven waiting for her under the covers in the big bed. Raven, who would sleep with his face pressed into her shoulder, his hand between her legs.

When she woke up he was gone.

About three weeks after the bread baking, Alice, on her own again, the girls in town with Peter, was sitting in the front room, thinking about taking a walk up to Harold and Stella's, when she felt an itching at her temples. She put her fingers up and pinched them together. A louse.

There was still time to catch the evening ferry so she jammed all her hair into an old beret and went to town. She did not know what you got for lice and it took her a while to find a drugstore that was open. The pharmacist did not laugh when she asked what to do but she could feel how hot her cheeks were, could imagine how red they must be.

She did not know whether to call Peter or not. Supposing he were with Anne-Marie? But what if the kids had lice? What if they infected the house in town?

In the end she didn't call; if they had lice it could wait until Sunday night. If they didn't, they didn't.

She called a friend and went there instead, washed her hair and dried it in front of the fireplace. Alone and miserable, curled up on a lumpy sofa, she wondered if there really was a last straw. The whole bloody cabin was probably infested with lice. Sheets, pillowcases, towels. And this was supposed to be her weekend off, her quiet time which relaxed and refreshed her, so that she would be on the dock on Sunday night all smiles and good humor. Good old mum. Keeping the home fires burning.

She couldn't sleep, thinking of all those lice. And hoped that Raven was itching too.

Harold and Stella came and got them one night. "Hurry up; it's beautiful." Harold held a kerosene lantern down over the government wharf. Soon the water was full of colorful worms, about six inches long, some half-green, some half-red, swimming rapidly near the surface of the water, darting in all directions. "Pile worms," Stella said, "one of the Indians told us about them, they're spawning."

"Why are some green and some red?"

"The ones with green behinds are male, the reds are female."

And then they stopped talking and just watched the show. Ripples and wiggles and dartings of iridescent color beneath the light from Harold's lantern. There was something almost terrifying about it, this ritual dance of speeding worms, the green posteriors male, the red female, shining just below the surface of the water. Later Stella said it reminded her of an orgasm, which she always thought of in terms of exploding colors. "Yes," Alice said, "and that's why it was so strange out there last night, all of us leaning over the edge, watching that intensity. I felt I'd gone back into some kind of knowledge which has nothing to do with our 'knowing,' nothing at all. And yet the links were there; sperms and eggs, sperms and eggs — we've got 'em too. And we too do our 'dance by the light of the moon.'"

"It's all sexuality, isn't it?"

"That would appear to be the case."

At the end of the summer, when Peter and Stella had got

together, Alice found a letter in the house in town. She had had letters from Stella so she recognized the handwriting.

"I'll meet you at the ferry," it said, "don't bother to wear any clothes." Showing his bright green behind to the other disembarking passengers.

Stella and Harold arrived at the door with an enormous roll of black plastic. Alice had invited them to dinner but Stella said no, she'd be left doing the dishes for six people and she hated doing dishes. However, they'd come over later on.

Harold told everybody to wait in the kitchen and then he and Stella took the plastic and some masking tape and went into the front room. After a while Harold called to Alice to turn off all the lights. Then they were invited to come forth. It was a moonlit night and they had no trouble crossing the original front room, where the fireplace was, but when they went into the new room it was terribly dark—all the windows had been covered over with black paper and the fire in the pot belly stove was out.

As soon as they were in, Stella sealed up the entrance as well. They sat on the rug in total darkness. Flora cuddled up to Alice; she didn't like the dark.

"Now what?" Anne said.

"Now we just sit for a while."

"I don't like it," Flora said. "Mum?"

"It's okay Flora; I'm right here."

There was no light. The room did not become lighter the longer they sat. Alice began to wonder about what was coming. The "dark" was never this dark—or not after one's eyes became used to it. There is dark and there is black. "Pitch dark."

Then Stella said, "Okay" and they could hear movement. A small sound.

Light flooded in through the pinprick Stella had made in the plastic. It was like a single star in a black sky. It was like a long narrow tube of light.

"Oh," Hannah said. "That's lovely."

Another light appeared. Then another, then another. Like a sky full of stars on a moonless night.

"Which one of you thought of this?" Alice said. Friends, children, bed, chairs, stove—all there, all visible now. What a relief.

"Harold," Stella said. "We tried it at our place first. Isn't it great?"

"Let's not turn the lights back on just yet," Hannah said. So they sat there, quietly absorbing the light, the relief from that utter darkness. Then Harold said, "Show over. Light fire now. Tea. Cookies."

"Can we do it for dad next weekend?" Anne said. "Is there enough plastic left?"

"I don't think so," Stella said, "but call him and have him bring some over."

Alice imagined them all, sitting there in the dark, Stella making the first pinprick, Peter delighted.

Later, sipping white wine in the Bengal Room of the Empress Hotel, Stella said, "The trouble with that year was there just weren't enough men to go around."

Alice couldn't help chattering on a little bit about poaching. But Stella hadn't really poached. It had all been over between Alice and Peter, over for months and months. It was unreasonable to expect someone like Stella not to flirt. Especially when she was getting bored and restless. Especially when Peter told her he felt no guilt.

Rivals are riverbank sharers.

Alice started making a list of people who had given their names to things. She called it "Household Words":

Lord Cardigan
Quisling

The Earl of Sandwich
L. Von Sacher Masoch
The Marquis de Sade
M. Guillotin
Mr. Condom (nationality unknown)
Mr. Hoyle

On weekdays Anne got up early every morning, lit the kitchen stove and made Alice a cup of coffee. Then they had breakfast. This was their quiet time together and they both liked it.

"It must be hard to be a 'middle,'" Alice said to her once.

"It's all right."

When Anne went down to catch the yellow school bus outside the store, Flora was allowed up. She was also allowed a very milky cup of coffee. After she went down to the baby-sitter's across the road, Alice made more coffee and sat with Hannah while she had her breakfast and then, one at either end of the small kitchen table, they began their day's work. It was a pleasant routine for all of them. Alice had never felt closer to her children. There seemed to be more time to talk to them, to play with them. It wasn't Peter's absence, really, although of course that had something to do with it. It seemed to be the island itself and the simple life they were leading. They got up early and went to bed early. They rarely used the car except to go to the ferry or on a picnic to the south end. They listened to the radio in the evenings, they played games, they carded wool, they read, they gave each other back rubs in front of the fire. The cabin lent itself to intimacy because there were no doors between any of the main rooms. The girls' bedroom had a cloth curtain at the entrance, that was all. Each one could feel the others' presences in the night. Sorority. In the best sense.

It was also to do with the wood fires and the candles they used as often as possible.

It was to do with walking up the road in a group to call

on Trudl and Christobel and Glenn or, further up, on Harold and on Stella. When the moon was bright they did not even use the lantern.

It was definitely to do with being out of the city.

Although her heart ached for Peter and she dreamed about him far too often, it was only when she was being melodramatic that she maintained he had "stuck her out here on an island" and then left her. He was really a family man. Whenever he had been away on workshops or conferences he had always written her letters which underlined this. He was not living with Anne-Marie but in a basement suite somewhere. Alice didn't want to know where, not even the phone number, in case she might call him up or go to see him in the middle of the night. They no longer, or rarely, since the black eye, met in town. He might be lonely even in his new freedom. She sent a grocery list in — of things she couldn't get — and he brought them out and left them when it was his turn to spend the weekend at the cabin. She often found an *Atlantic Monthly* or *Canadian Forum* left as well. Sometimes a fish, cleaned, in the refrigerator. It seemed to her that he had left her, yes, but with love, not with indifference. He might still come back. Meanwhile she must get on with her life.

So Alice had felt in her more optimistic, usually-in-the-daytime, moments.

In the coffee shop Peter was drawing with a red pen on little pieces of card. They had been figuring out some expenses but that was finished and still he ordered another two coffees. He drew over and over, with elaborate shading, the numbers "1," "2," "3." Gave her his little smile and Alice went cold.

"Now don't go all electric on me," he said, "but I have something I want to tell you."

Alice stirred her coffee.

"Oh yes?"

The waitresses in the Bavarian Room wore dirndl skirts

and peasant blouses. Their waitress's name was Ursula. We always go to restaurants and coffee shops, she thought, when he has anything big to tell me. So I won't make a scene. In nice ethnic surroundings. The Hong Kong Kitchen, the El Matador, Lindy's Kosher Delicatessen and Restaurant, now the Bavarian Room. How do you say "pain" in Bavaria?

"There's a possibility I may be offered a job in the north of England," he told the placemat. "An exchange with a fellow in Newcastle-on-Tyne."

"I met somebody from Newcastle once," she said. "I could hardly understand a word he said."

"Yes. Well. Of course I'd take the children, it being my year."

"Take the children."

"Of course."

"They'd be so far away." Alice was cold all over and she felt very queer. She forced herself to stay in her chair.

"That was the agreement."

"But we agreed to stay close!"

"It would only be for a year. Maybe you could get a charter flight at Christmas." Drawing on the placemat:

1 , 2 , 3.

1 , 2 , 3.

"That's not quite all of it," he said, "and anyway, nothing is settled."

1

"I'd like

2

to ask

3

Anne-Marie to go with me.

(pause)

Or I think I would, I'm not sure. But it's definitely a possibility. Certain other . . . things . . . have come up here. Certain other relationships are opening up so I don't know. I did feel it was only fair to tell you."

"Thank you," Alice said miserably. She didn't want *that woman* to live with her children. She didn't want Jeannie hitting Flora over the head with blocks because she was jealous of her. Oh oh oh, she mourned in her heart, what had she agreed to? How could she, how would she, bear it?

''It would be good for them,'' he said, ''and they're part English after all.'' Then, putting his hand on hers, which had turned to ice, ''I think it might be good for all of us.''

She tried to make a joke of it as he let her out to catch her bus to the depot:

''Hey!'' Leaning back through the car window.

He smiled at her, obviously feeling the interview had gone well. On his way to Anne-Marie's to offer her romance in the border counties.

''Hello there.''

''If Anne-Marie says no, how about taking me?'' She felt as though she had swallowed stones. But he just smiled and drove away.

And, in the end, he decided not to go.

Alice wrote to Selene:

Raven's been visiting us. The children love it because now they can fart and belch with impunity. But he's also had everybody out cleaning up the yard and has made us a beautiful wooden bowl. You're all so clever with your hands I'm jealous! He loves you Selene, you know that don't you? Even when he hurts you. He says he gets ''frantic'' at your goodness. Well, I can understand that. I used to be terribly jealous of you, the ease with which you could love people, could accept them. I'm not any longer. It seems to me that you have got to be a little less loving and forgiving if you are to survive. Your leaving for New York was a step in the right direction. I'm sure you are lonely but now Raven can see just how much he loves you. Don't stay away too long. I love the thought of you working in *Schraffts*! Do you wear a little uniform, like a French maid? I remember going there once, years ago. All the rich ladies were sitting around eating equally rich desserts!

I love you Selene. Knowing you has made me grow. I want you to be part of my life always. Funny how close I am getting to women — Stella, Trudl, you. I guess I always told my secrets to Peter before, except for my brief friendship with Anne-Marie. And to her I told too many!!

Take care — come back to us soon.

love,
Alice

Trudl and Glenn had offered to do the washing up. Alice was in the new room about to play Let's go on a lion hunt. She went into the kitchen to get a drink of water for Christobel. Trudl and her lover were standing facing one another, arms around each other's waists. Smiling.

''I'm sorry,'' Alice said. ''I didn't mean to interrupt.'' She got the water and went back to the other room.

''Now remember,'' she said to the children, ''you have to do everything I do, say everything I say.'' She sat down cross-legged and began to pound the carpet, alternately, with the palms of her hands.

''Let's go on a Lion Hunt.''

''Let's go on a Lion Hunt.''

''All right.''

''All right.''

''Let's go.''

''Let's go.''

The hunters moved through the jungle in their heavy boots. Alice tried not to think about the couple in the kitchen.

''Oh Look!''

''Oh Look!''

''There's a Swamp.''

''There's a Swamp.''

''Can't go under it.''

''Can't go under it.''

''Can't go over it.''

''Can't go over it.''

"Gotta go Through it."
"Gotta go Through it."
"All right."
"All right."
"Let's go."
"Let's go."

Desperately they pushed their way through the evil-smelling swamp. Mosquitoes stung their faces and bare arms, the mud sucked at their feet, but they pressed on. Gotta go through it. Gotta go through it. Gotta go through it.

("You *know* I love making love to you," Peter had said, "you know that. But I mustn't do it anymore; I have other commitments." She spent her last night with him trying to memorize his body with her fingertips.)

FEBRUARY

I slept with Peter last night—the first time in months. He was waiting to get on the ferry as I walked off. I said, "Peter, come back up with me, I want to talk. I'll drive you to the ferry in the morning."

But we didn't talk —or not much. I made hot rums and we sat on the floor by the stove. The girls were asleep.

He said how lovely the cabin looked and how well he thought the girls were doing. I told him a couple of funny stories and then I said, "Peter would you like to make love?"

"All right," he said. It wasn't much but it wasn't no.

I made us a bed on the floor, so that we would get all the warmth from the stove, and then I rubbed his back with some almond oil and then we started making love, in the dark, with just the light from the stove. I said nothing; I didn't cry; I just tried to tell him, with my body, how much I've missed him and how nice it felt to have him in me. Afterwards we got into bed and I fell asleep with my hand curled around his balls. They felt soft and warm and fragile, like the bellies of kittens. Little animals soft and warm in my cupped hand.

In the morning he was up very early, making the fire in both stoves, making coffee. I told the girls to go ahead and have their breakfast, I had to take him to the ferry. Flora was the only one who seemed genuinely glad to see him. They have a routine now: they see him every two weeks but never with me; we had upset that routine and it made them uneasy.

I didn't care I didn't care I didn't care.

At the ferry he ruffled my hair and kissed my forehead (he who had had his head between my legs the night before!) and went away.

I didn't tell him I had seen Anne-Marie, that I knew he was courting Penny now. Ron had been one of Anne-Marie's "husbands" and then went on to someone else. Penny is taking a class from Peter; now he drops by her house which is right around the corner from Anne-Marie's. A-M has a dog now, a puppy, and when she walks it late at night she sees his car parked outside Penny's house.

This knowledge gave me courage, the courage to ask Peter to stay last night. His "once in a lifetime love" with Anne-Marie is coming to a close. She says she still sees him but not so often anymore. She doesn't seem heartbroken about it—perhaps she has someone new. I couldn't compete against the glamor of Anne-Marie but Penny is ordinary, like me. Funny, generous, harassed mother.

And so I swallowed my pride (what pride? There couldn't have been much; it went down so easily).

I drove back up the island singing. The girls have asked

me no questions and I have volunteered nothing. This afternoon Hannah found some snowdrops by the shed.

"I think," Alice said, "that the minute I became a mother he was unable to love me any more. Romantically, I mean."

"You mean because you lost your figure?" Of course Stella would ask that.

"Oh I didn't, not for years. No, nothing like that. I mean because he hated his own — *she* was so uptight about sex. Mothers and fathers shouldn't be genitally oriented. Got to get on with it. Life, don't you know. First act's over, now we're into the heavy stuff."

"You hated your father."

"I didn't *hate* my father, I just found him repulsive. I hated my mother; used to — not any more, but you're right, the same thing can apply. Perhaps I did it to him as well. How sad."

"Squeezed by the head through a narrow hole. No wonder we're all crazy," Alice said. "The baby's down there trapped in a cave."

Trudl said: "I wonder if it makes a difference if you're a Caesarian? If you don't undergo that journey or come out the regular way."

"It must, mustn't it? I mean what an unnatural act. You're just lying there and then suddenly you're lifted up into the air, mother unconscious with a great wound in her belly. No immediate hugs. No warnings."

"Maybe that's what made Caesar the way he was."

"?"

"He still wanted to emerge triumphant."

"Look what I brought back from town," Alice said, "yerba maté." They were all reading Cortazar's *Hopscotch*.

''*The* yerba maté?''

''I hope so. Let's try it.'' She put the kettle on and they waited.

''How much should I put in?''

''Doesn't it say?''

''No.'' Alice put in a big handful and then strained it through her wicker strainer. They sat over their cups of yerba maté. Stella spoke first.

''Are you enjoying this?''

''Are you?''

''Not much.''

''It's terrible.''

''Awful.''

''Maybe we need to make it in a gourd.''

''Maybe you put in too much yerba.'' Hannah came in the kitchen.

''What are you two drinking?''

''Yerba maté.''

''What?''

''Yerba maté. It's a South American drink. The people in *Hopscotch* drink it all the time.''

''You two are crazy, you know that, don't you?'' She tasted it. ''It's horrible.''

''If Julio likes it, we'll learn to like it.''

''Or die trying.''

''Our pitiless bottom,'' Alice said,

''Our bottomless pit.''

''Yes,'' Stella said, sipping her yerba maté (they were getting used to it now, cheering each other on), ''but can you imagine being old and not wanting it?''

''Do you think you ever really don't want it? It's probably just that the old ladies dry up and the old men get limp so the whole thing becomes too sad and painful, too humiliating to one or the other or both.''

''I just don't want to think about it, it makes me shudder.''

''You know,'' the lawyer said, ''I'm supposed to counsel you if there's any hope of you two getting back together.''

Alice waited for Peter to speak. She looked out the window and tried not to cry.

''Mrs. Hoyle?''

''There's no hope.''

''It can always be rescinded, you know,'' the lawyer said gently. His name was Mr. Precious; Alice had never seen that name before. She'd have to add it to her list. ''If you live together for thirty days the agreement is revoked.'' They told him they wanted to compose their own agreement and then he could turn it into legalese.

Before Peter dropped Alice off at the house she asked him to stop at a Chinese grocer's. She bought him a big bunch of chrysanthemums, the bronze ones. She had dressed up for the occasion in her long tweed skirt and the brown cape she had made. She knew she looked very nice.

''Here,'' she said, ''these are for you. A new blooming.'' He had often left her flowers on the table at the house.

''They're lovely.'' She was bleeding everywhere, couldn't he tell? Like the Little Mermaid, she would leave a trail of blood.

They kissed good-bye at the front door.

''I'll see you soon. I've got to catch the ferry.''

She went inside and sat very still for a long time, afraid to move. A line from *The Rise and Fall of Arturo Ui* came into her head. She had been to see it the week before. The line where the girl runs onto the stage, badly wounded, and says to the audience,

''Give me a bandage or I'll bleed to death.''

''Unhappy differences have arisen — ''

Alice noticed that the document said "Peter Hoyle, teacher" and underneath, "Alice Hoyle, housewife." She borrowed one of Hannah's felt pens and crossed that out, that housewife business. "Writer." Then she wrote a letter to the lawyer and enclosed the agreement saying that she had no intention of signing until the document had been officially changed.

She had heard that the Duke of Windsor always put "nothing" under "occupation" on his passport after he stepped down from the throne. She wondered what Mrs. Simpson put. "Companion to ex-king?" Perhaps she should have crossed out "housewife" and put "castaway" or "reject" or even "lunatic, part-time."

"Well I know *I* could never share," Stella said.

FEBRUARY

Deliver. v.t. Rescue, save and free *from;* disburden (woman in parturition) *of* child (usu. pass; also fig. *was* ~ed of a Sonnet); unburden (*of* esp. a long-suppressed opinion etc.) in discourse; give *up* or *over,* abandon, resign, hand on *to* another; distribute (letters, parcels, ordered goods) to addressee or purchaser (~ *the goods,* fig. carry out one's part of agreement); present, render, (account); (Law) hand over formally (esp. sealed deed to grantee, so *seal & ~*); launch, aim (blow, ball, attack; *battle,* accept opportunity of engaging);

recite (*well ~ed sermon*). Hence ~ABLE a. (f. F *deliverer* f. LL *deliberare* (DE-, L *liberare* f. *liber* free)

Let no man put ass under.

Sometimes when it rained all weekend they got out the jigsaw puzzles. The month before she had gone to Woodwards to get some new ones, was amused by the banality of the titles but too tired to go to a classier place.

"Church Scene in Bavaria"
"Farm Scene in 60 Pieces"
"Newfoundland/Terre Neuve"
"Maple Forest, Shawand County, Wisconsin"
"Vancouver, BC in 750 Pieces" (Also Jasper, Montreal, Toronto, Notre Dame Bay)
"Donald Duck"
"Goofy"
"Home Sweet Home" (an imitation sampler)

Finally, way in the back, she found a very nice zodiac and heaved a sigh of relief. Waiting in line to pay, she thought of another title:

"Alice Hoyle: 1,000 Interlocking Pieces."

Friends, a little bird said to me, those people would like a toke, so i walked over with my joint and offered them a puff. They gratefully inhaled deeply, not wasting any smoke, holding their breath, a power inside welled up and from their souls, love, emitted.

today i returned home from a hike on the Mountain. i climbed till i was knee deep in snow, then i crossed over to a less snowy ridge where the sun got to fall a lot more, climbing to the top, peeking over the edge, to my eyes did appear a lake so beautiful i didn't want to leave, but nite was coming, the Mountain didn't have too good a wood supply so i followed a river further down to where a big fir tree grew, i lit a fire, baked a sweet potato, and settled down for a good night's sleep. it rained during the night, but the trees sheltered me and only a few fine drops got into my bed. a fire in the morning dried my being and belongings and cooked a pot of buckwheat to which i added raisins and nutritional yeast then onto the trail home.

the letter you sent prompted a spirit to show me a small stash which i now pass on to you with love

your brother in god —

Raven

xoxoxoxoxoxoxoxoxox

p.s. the piece of paper is enclosed to help conceal bulky luggage and to inform you all that

I LOVE YOU ALL

try to puff together, you get higher.

FEBRUARY

Cock (1), a male bird. (E) M.E. *cok*. A.S. *cocc;* from the bird's cry. 'Cryde anon cok! cok!' ch. C.T. Nun's Priest's Tale, 457. cf. Skt. *kukkuta,* a cock; Malay *kukuk,* crowing of cocks. And cf. *Cuckoo*.

cock, the stopcock of a barrel, is the same word.
cock, part of the lock of a gun. From its original shape; cf. G. *den Habn Spannan,* to cock a gun.
cockade, a knot of ribbon on a hat.

cockerel, a young cock.
cockloft, upper loft.

Cock (2), a pile of hay. (Scand.) Dan. *kok*, a heap; prov. Dan. *kok*, a hay cock, *at kokka höet*, to cock hay; Icel. *kökkr*, lump, ball; Swed. *koka*, clod of earth.

Cock (3), to stick up abruptly. (E.) Apparently with reference to the *cock's* head when crowing; or to that of his crest or tail.

Hen. (E.) A.S. henn, hen, han; a fem. form (Teut, type *han-ja) from A.S. *hana*, a cock.

"You know when I think I fell in love with Peter?" Alice said.

"When?"

"We went to a Frankenstein movie one day in Birmingham. I got quite frightened and Peter said, 'Close your eyes, I'll tell you when to look.' I said okay and there I was, with my eyes closed and he was saying, 'Don't look, don't look, don't look.' And all of a sudden I peeked and he had his eyes closed as well!"

She sighed. "I'm sure he doesn't remember that — he has a very bad memory — but last October he said to me, 'You don't want a husband, you want a seeing-eye dog.' A very strange remark and I've been puzzling over it ever since."

"He probably meant you want a protector, something as simple as that."

"I don't know. With a blind person and a seeing-eye dog who's the master and who's the slave? The dog may be 'protector' but it is also totally bound to its master's wishes. I doubt if they ever get time off, take a holiday etcetera. I think that's more what Peter meant. That I relied on him absolutely—that he would always be there, totally bound to me and my wishes."

"Weren't you always there for him?"

"Probably not. The children, my work. Artists are never 'always there' and mothers generally put children before husbands. Husbands are supposed to be strong and able to take what comes. I assumed he was all right. He had to be. I needed him to be—there was always so much going on."

Stella stubbed out her cigarette and lit another.

"I wonder how many seeing-eye dogs run away?"

"I wonder."

"Women aren't encouraged to take risks," Alice said, "except to risk their lives to have a child."

"They're taking risks now." Trudl helped herself to another brownie.

"How do you stay so thin and eat like that?"

"I don't know, but I'm glad."

"I was skinny, well, no, slim, until about four or five years ago," Alice said.

"Well your metabolism changes after thirty, I guess. And think of the way we ate as teenagers. God! That amount of food would disgust me now."

"I was always hungry," Stella said. "For one thing or another."

They all laughed.

"Well, it was food with me, not sex," Alice said. "I don't think I felt any specific sexual feelings until I was about seventeen. I got *crushes*, lots of them, but it was all very romantic. Saw myself being rescued from a burning building by my math teacher, stuff like that. I never actually thought about anyone *fucking* me. Typical late-forties, early-fifties teenager. Although it may have been because boys didn't really like me."

''I was afraid of boys even though they liked me,'' Trudl said.

''Why?''

''Afraid of my feelings, I guess.'' Trudl in a white wedding dress, her red hair streaming behind her, walking the two blocks to the church with her father. Alice had seen the photo album.

''You should have seen them,'' Stella said. ''Trudl and Jimmy at Halfmoon Bay. Falling on each other's necks when he came home from work. Adoring Christobel.''

''What happened?''

''She started taking painting lessons and fell in love with her teacher.''

''And?''

''And.''

Alice remembered that painting Trudl had done just after she left her husband. A young woman with red hair crawling out of an enormous box.

After Trudl came to the island, or perhaps after she fell in love with Glenn, she went through her misty phase — all swirls and chiffon scarves. She gave Alice and the girls one of these misty paintings for Christmas. Alice had liked the girl coming out of the box much better. Trudl and Glenn played their guitars very softly, sang together softly and made soft eyes at one another.

When Stella told her mother that Glenn had moved in with Trudl and wasn't it romantic, Stella's mother said:

''Wait until she has to do his socks.''

''I never fucked anyone until I was nineteen,'' Alice said. ''I wanted to but nice boys didn't do that or not to nice girls. I didn't know any nasty boys and used to wish I did. The big tough guys at the insane asylum where I worked in the summers used to say things but they were all afraid of me. I went out with one once, out to a drive-in. I remember feeling 'This is the night,' so I took the stuffing out of my bra when we went for hamburgers and chips. He had a blanket in the backseat, I could see it. But then, when we got out to some field he knew and were lying on the blanket I told him I was a virgin. He stopped dead. Said he couldn't do it to me. He disappeared for a few minutes then we got in the car and drove home."

"You'd think he might like it if you were a virgin."

"Maybe he was Catholic — most of the people who worked in that hospital were, I don't know why. Or maybe the word had too many connotations."

"I started fucking when I was thirteen," Stella said. She grinned. "I wish I'd started sooner." Stella and her girl friend, later, drove around in a car and picked up boys from senior-high-school playgrounds. Alice could not imagine being that sure of herself.

Trudl had never tried oral sex until she left Jimmy.

"Do you like it?"

"It's different."

"Better," Stella said, "much better."

And when Peter started suddenly liking it she should have known.

"Open your legs," he said. Kneeling at the foot of the bed, his tongue exploring her. She loved it — it drove her wild. She nearly suffocated him, pushing up against him — Oh Peter I love you don't stop don't stop don't stop.

And she, sucking him, her face in his belly.

"You come too quickly that way," he said, "why not take your time?"

"I can't help it." She laughed. "I love it."

"You come too quickly that way." In relation to whom?

Later he told her he was impotent with Anne-Marie.

"Have I done that to you!"

"I don't know."

She tried to imagine it but couldn't. Did she suck him and touch him? Did she stand in her beautiful nakedness in front of her little fire? How could he not? Was that why he had left? Because Alice was controlling him somehow, or making him that guilty? And yet, months after Peter's confession, months after he had left, Anne-Marie said to Alice,

"He's always been impotent with me."

No wonder, in the end, he went away from Anne-Marie as well.

Alice and Stella were talking about writing.

"Oh, I don't know," Alice said to Stella, "it's a funny feeling—like knowing just when the bread has risen enough—or maybe even baked enough. A kind of 'knowing' anyway, that you're doing it right." Then she laughed.

"But sometimes that's an illusion. You burn it or it's underdone or soggy — a mess."

"And then?"

"And then I crumple it all up and use it to start the fire."

Sometimes Alice saw, growing between her and the other three women, a great twisted vine, or rope. So that, if she had had to step out into the dark, she could, as she had once done on a small ship caught in the tail end of a hurricane, pull herself forward safely, even in the most severe of storms. Could move from her cabin, to Trudl's, to Stella's, even over to Selene's in the most awful storm of blind despair or self-hatred and know that so long as she held on tightly to their friendship she would be *all right*. That there would be warmth and light and a change of clothes and something hot to drink.

She shared her image, separately, with her friends.

"That's beautiful," they said, and yes, a rope. Perfect. Alice with tears in her eyes. But smiling. The hell with seeing-eye dogs.

FEBRUARY

TIGER BITES TRAINER TO DEATH
(HORRIFIED WIFE LOOKS ON)

YOUTH KILLED IN RITE

Was it? Was that what happened? ''Marry in haste, repent in leisure.'' Youth killed in rite. Our youth. Was our marriage (L. *maritus*, a husband) a sham? I'll never know. If he thought it was then it was—for him. He wanted me to be more sexy, bought me lovely nightgowns and negligees which I hardly ever wore for fear of setting myself on fire. I liked the coziness of flannel, especially in the middle of the night, getting up to soothe or feed a child. When miniskirts were in he bought me miniskirts and go-go boots. I thought that was kind of fun but felt awkward sitting down or bending over. Once or twice I found porn novels under his pillow. I ignored all these signs, these signals from a man who found it almost impossible to articulate his feelings. He took on more on more jobs, workshops, life classes, children's Saturday classes. Sometimes he went for a drive in the car, by himself; sometimes he went to the movies. I stayed home and cooked and read the children fairy tales. I wrote at the kitchen table. I (slowly) got a degree.

Farther and farther apart.

Water imagery again—I can't escape it. ''Your honor, we gradually drifted apart.'' Each in our little boat, giving the occasional wave or halloo. How sad. That's what it really is; it's sad. When I'm not full of rage I'm sad. But I mustn't look back so much or my tears alone will turn me into a pillar of salt.

He never beat me; he never laid a hand on me except

that once, when he punched me in the face and then came to tell me he loved me. I'm still puzzling over that one.

"Love" — "in scoring in various games, as tennis, rackets etc: No score, nothing; *l. all*, no score on either side. 1742."

"Alice what do you really want."

"Nothing," I said, "nothing." When what I meant was "love."

Peter was now going around with Penny.

"How's Penny?" Alice said.

"Do you really want to know?"

"Yes, I guess I do. I know she's been through a bad time."

"Well, she's all right now." He gave her his little smile. "I feel that every time I tell you something like this," he said, "that I'm just putting another nail in your coffin."

"That's a pretty horrible image. Do you hate me that much?"

"No, no of course not. I didn't mean it that way."

"How do you mean it?"

"I don't know. Alice, you're trying to pick a fight."

"No I'm not. You see me in a coffin and yourself as nailing me in and you say it doesn't mean anything. It's a lovely Faulknerian image, southern-Gothic. They don't, in this country anyway, nail people into coffins any more."

"Let's drop it."

"All right. Let's drop the coffin. Let the dead carry the dead. I've got to go."

She stood up.

"What a pity it is," she said, "that husbands can't put

down obsolete wives the way one puts down a superannuated cat or dog."

"I'll get the stuff out of the fridge," he said.

"You do that."

"Have you ever made love to a woman?" Trudl said.

Alice hesitated.

"Once. With Selene."

"Did you like it?"

"'Like' isn't the right word. It was a strange experience for me. I was unutterably sad and longed for Peter to come back. She was sad too. Raven was treating her very badly and she knew she was going to have to leave him. She was sleeping on the floor, by the windows, and one of the top windows was open a little bit. I was wide awake in the big bed and looking out at the moon. This phrase kept going through my head. 'The white wrinkled belly of the moon.' Giving birth to so many nights, you know. Or so many days, depending on how you looked at it. One of the children had a cold and was sleeping in my bed so I was lying very still, looking at the moon, thinking about Peter, praying he would return, beginning to realize in my heart, perhaps, that he never would.

"And then I realized Selene was awake too, lying on her face and weeping very quietly into her pillow."

Alice paused again, trying to explain. Trudl sat very quiet, waiting.

"I think it was really the *mother* in me that made me get up. Although Peter always maintains that Selene is so centered, she seems to me to be an incredibly lonely person. She denies whole continents of herself in order to keep loving Raven. *She* is always the calm one, the peacemaker, the 'parent' in that relationship. I just wanted to hold her while she cried — the way you do a child.

"So — I didn't say anything but slipped out of bed and just lay down beside her and started rubbing her back. She was the one who taught us all about how lovely that can be.

Then she whispered, 'You'll catch cold,' and turned over to put her blanket over both of us. Her face was covered in tears and shone in the light from the window. Like porcelain.''

''She has a lovely face.''

''Yes she does. Like a saint or an innocent, although she is neither. Anyway, I lay on my side and started rubbing her neck, her breasts—she has a lovely body too, and she began to touch me. Then she turned toward me and buried her face in my neck and whispered, 'Oh, you smell so good,' just like a child does sometimes, and I started making love to her, and she to me. It wasn't erotic. I don't know what it was. Some kind of strange communion.''

''Did you come?''

''Yes. We both did. But not passionately. I can *not* explain. We knew it would never happen again but we were both glad it had happened. What's that book by Radclyffe Hall? *Well of Loneliness*? I guess that's where we both were. Down in that well with the moonlight streaming in upon us. It was like a dream. We met on this deep plane of utter despair. It was—it was a conversation.''

''A conversation?''

''Yes.''

''Sometimes it can be like that with a man.''

''Yes it can. But when it's a man and a woman, or I guess a 'couple' in any sense where lust is involved, something often goes wrong. I wanted her and Raven to work things out; she wanted the same for me and Peter. As I say, it was a very strange thing.''

''Had you ever been erotically attracted to a woman before?''

''Just preteen crushes on camp counselors at scout camp. I've always been afraid of women, although—''

''What?''

''One day last spring Anne-Marie mentioned on the phone that she had some psilocybin and would love for the three of us to go out in the woods somewhere and take it. You know how scared of drugs I am. And yet I was tempted.

"But I said, 'Then we could both make love to Peter' and was surprised at myself."

"What did she say?"

"She said, 'Why not?' Of course they were seeing each other secretly by this time only I didn't know it, or wouldn't know it. But what is important here is that I added, 'Well, why not is that I guess, to complete the experience, you and I would have to make love as well.'"

"What did *she* say to that?"

"She laughed and said, 'Touché' or something or 'I suppose you're right,' and that was the last I heard of it.

"They probably took it anyway. She was always offering him dope or hash or something. Perhaps I *was* attracted to her, I don't know. I've realized this year how little I know about myself, let alone other people, even people I deeply love. I know I was, consciously, afraid of her and at the same time contemptuous of her.

"But I hadn't been alone, then," Alice said. "I didn't know about all those dark watches of the night. How the loneliness wears you down. How desperate you feel, how ugly even if you aren't, how left out. I only saw the beauty and the wasted talent and heard the whining and self-pity.

"I remember Stella describing coming downstairs just after Robert had died and seeing the friends she was staying with embracing. She said she wanted to kill them. 'Murderous rage,' I think she said."

"I've got to go," Trudl said. "I promised Glenn a game of chess. And then we're all going to go and have baths at Stella's dad's before it gets too late."

"See you soon."

"No doubt tomorrow night."

"No doubt."

Alice sat at the kitchen table staring at the candle and the teacups. She had been laying out a pattern for a summer nightgown when Trudl came. The first of three, at least they didn't have sleeves. Why couldn't she be like Stella and not just admit she couldn't sew but genuinely not care whether she could sew or not? Never mind, and Trudl had said she'd come to the rescue if necessary.

Selene sewed without patterns and by hand. Beautiful as well as useful things. So did Anne-Marie.

The Scorpio Housewife gathered up her things and went to bed.

And realized just before falling asleep that she hadn't asked Trudl whether *she'd* ever made love to a woman. And sleepily wondered why Trudl had brought the subject up.

"Women have been shanghaied," Alice said, "and now we are waking up and rubbing our eyes and murmuring, 'Where are we?'"

"What's the answer then?"

"Some kind of mutiny, I suppose. Unless we can talk the captain into letting us go."

"But that's where your argument falls down," Stella said. "We don't want to be 'let go.'"

"True, o queen. No, we don't want to be let out at the nearest port or unceremoniously tossed overboard—although that's the fate awaiting a lot of us I think — what we really want is to be officers and captains ourselves. It's funny, since I've been reading all this history about the Spanish and the English and the Pacific I think more and more in maritime images. I'm like John Donne in his love poetry. Navigational instruments, new lands, maps, merchant ships. His language reflects what was going on around him in the outside world. I read about prizes and shipwrecks and plunder, strange instruments which measure the artificial horizon, about conquests and conventions, this whole male world of the age of exploration and I see that women are going to have to get out there and do the same thing."

"Conquest and plunder?"

"Well no. No. We don't want to emulate the male with

all his bad habits, now do we? What I mean is that what's happening to men and women today is just as exciting and terrifying as the discovery that the earth was round, not flat, or even that the earth was not the center of the universe but just part of a solar system. But we all need new maps, new instruments to try and fix our new positions, unless we think we're competent enough to try and steer by the stars.

"And to go on with my maritime theme, imperialism is over, for nations, for men. Do we really want somebody planting a flag on us and claiming us forever?" She laughed.

"I talk big, of course, but I guess that's maybe what in our hearts we still want. I was looking up 'abandon' the other day and discovered it, literally, means 'to set at liberty.' I say Peter abandoned me, and mean 'poor me,' when maybe I should be feeling 'he has set me free.' "

Stella lit a cigarette. "But what you really feel is that he has tossed you overboard."

"Yes."

"We all want one special relationship," Trudl said. "We're all looking for Prince Charming."

"Yes," Alice said, "and we ought to be ashamed of ourselves."

Trudl looked worried. "Don't you think it's natural to want to be loved?"

"Absolutely," Alice said. "And to be loved deeply and intensely. But we can't make that our whole life, our whole reason for being. Women have *let* men define them, taken their *names* even, with marriage, just like a conquered or newly settled region, *British* Columbia, *British* Guiana, *New* Orleans, *New* Jersey, *New* France, *New* England, etcetera. I really understand all those African nations taking new names with their independence, names that relate to their racial history. Also the Afro-Americans, and the Indians."

"But what could women change their names to?" Stella asked. "Isn't it really as pretentious if we do that as all these hippie girls calling themselves Rosamund Apple or Johanna New Moon or whatever? I mean, we can take back our *maiden* names, such a quaint term, like 'horsepower' or one of those isn't it, but they were given us, we didn't have any choice. Or about our first names.''

''Well, perhaps we have to change all our names,'' Alice said, ''I don't know. Somehow one's first name doesn't seem as politically oppressive as one's married name. And yet I loved being 'Mrs. Hoyle.' Then everybody knew I was married, you see. I had status. I had an official lover and protector. You'd go to a party: 'I'd like you to meet Peter and Alice Hoyle.' A couple, a unit. I loved all that stuff.''

''I did too,'' Trudl said, ''until I wanted to leave.''

''I never wanted that,'' Stella said.

''Yes, but we all know you're the miraculous exception. And I'll bet you would've wanted it with Robert if you hadn't met him when he was already dying.''

''Maybe.''

''Are you going to go back to your maiden name?'' Alice asked.

''Müller-Stach? That's quite a mouthful, isn't it? I don't know.''

''It would be nice if one could just be 'Colette' or 'Garbo.'''

''I guess. I've actually been thinking of changing my name,'' Alice said, ''but the children seem frightened by the idea.''

''Why?''

''I'm not sure, since they call me mom or mummy anyway.''

''What are you thinking of changing it to?''

''I don't know, that's the trouble. I like 'Justine' but of course *that's* out. I hardly look a neurasthenic Jewess. I've been trying to think of something Irish but not sentimental. Nora, maybe, d'you like that?''

''No. Not really.''

''What about Suzanne?''

''Yes, that's pretty and a nice bilingual touch. But it doesn't suit me.''

''What about Christine?'' said Trudl. Stella said, ''What about Pristeen?''

''Del-feen.''

''Vaseline.''

''I'm serious,'' Alice said, laughing. ''I need help.''

''We're helping you.''

''Thanks a lot.''

''Women and men make terrible bargains with one another,'' Alice said. 'I'll be mummy if you'll be daddy.'''

''That's changing.''

''Yes and no. What are the hippies doing over at Coon Bay? The women cook and have babies, follow the men around, do the washing, all that.''

''But the men don't earn money.''

''That's true. Often the baby is the key to the welfare check, right? The key to his month's grub and dope.''

''And the men can always split if they find a new old lady.''

''God, is there any hope?''

''I don't know,'' Alice said, ''I really don't. I used to think Raven and Selene were the perfect couple—living off the land, never raising their voices, moving with the changes, never eating anything that died in pain. But they have the same fuck-ups as the rest of us. Raven treats Selene horribly sometimes, maybe often, and she comes back for more. Won't even let her have curtains in the new cabin—'only essentials,' he says.''

''She doesn't have to stay.''

''She loves him.''

''Whatever that means.''

''Yes.''

It was nearly midnight. All the children were asleep.

''The woman artist has an even harder time,'' Alice said. ''If she is to move forward at all she has to develop a layer of selfishness — self-is-ness — that has been traditionally reserved for men. But then the men, who feel 'despisèd and rejected,' accuse these women of being 'cold.'

'''If you love me why do you neglect me?' or words to that effect.

"As though all the heat, all the energy of the woman had to be directed toward the man!

"And the woman, guilty in her *own* eyes, never mind the man's, weeps and asks forgiveness and accepts the man's point of view as the correct one—or the only one. She backs down and puts the dinner on. Or else.

"It's like what happened to Peter and me. He encourages me and encourages and encourages me and then, when I'm really committed to writing, really flying around out of the nest a bit, he says, 'You're cold, you're calling the shots, you're neglecting me.' That's an oversimplification of course but it's not far off. The stronger I got the more he turned away from me. Poor Anne-Marie, she needs a shoulder to weep on, kind of thing. Then tells me I'm 'emotionally dependent' when I'm heartbroken that he's leaving me."

Stella was rolling a cigarette.

"Would you really have shared?"

"I think so; I don't know. Certainly for that year he was remaining in the city and only coming out on weekends. If we were all together I don't know."

"I could never share," Stella said, "never."

"I probably only could in retrospect."

"I mean, how would you work out the practicalities," Stella said. "'It's your night dear.'"

"I don't think I could've done it under the same roof. Or even in the same town."

"I'd tear her eyes out," Stella said.

"I'll bet you would."

"Trudl and Glenn locked me out of the cabin on Sunday," Stella said.

"*Your* cabin?"

"Yes. Christobel was in town and Harold and I had gone to dad's to take a bath. We came back and the door was locked."

"What did you do?"

"We snuck around and banged on all the windows. I hope he lost his erection."

"I can't imagine Glenn fucking."

"But can you imagine fucking Glenn?"

"In an abstract way."

"Like fucking an angel maybe."

"I'd be afraid I might damage him, he's so gentle."

"It's hard for me to say. I've had incestuous yearnings from time to time, but then I look at him and wonder what all the fuss is about. 'Same old Glenn.' It's pretty hard to keep him on that exalted plane when you've seen him with his legs in the air, having his diapers changed. Listen," Stella said, "will you walk me partway up the road? It's awfully dark out there."

"Of course. Or you can take Byron. He can take his toothbrush and spend the night."

"Thanks," she said, "it'll be company for Uggah. I'll bring him back when I come to the store in the morning."

After Stella left Alice threw open two of the windows to get rid of the cigarette smoke. She liked Stella, liked talking to her, looking at her. Liked her better than she had ever liked Anne-Marie. Liked her better than Trudl, in spite of the mother bond. Trudl wasn't real most of the time. She had just exchanged one kind of box for another.

Stella's smile was beautiful because it was full of pain.

Stella showed Alice an entry in her journal:

"We are born before we fuck."

"Meaningless, right?" she said. "But I believed I was trying to say something at the time."

Alice leaned out the window in her nightdress. The night rustled around her like dark silk petticoats. Just when you think you are most alone someone comes along with whom you can really connect. "Only connect" swam up from the aquarium of her mind. Who said that?

The night smelled good. Slightly damp, fertile, a warm, deep-brown smell of leaves and sea. She wondered what Peter was doing but the pain was lessened by the evening she had spent with Stella.

"We're tough," she thought, "Stella and I. Life deals us cruel hands but we keep playing."

She closed the windows and went to bed.

In the middle of the night she was awakened by the smart click of a mousetrap in the store cupboard.

Then silence.

She turned over and sighed and sank back down in her dream.

"I remember a woman whose husband had left her saying to me, 'Well at least you have the children,' but of course the really terrible thing I had to face was that Peter *wanted* the children, would have taken custody of them gladly. It was only me he didn't want. That's pretty awful."

"Could you have given them up?"

"I thought about it—for all our sakes. He's a very good parent, he can make more money than I can, he's not neurotic and insecure. The leaver is always in a stronger psychological position, don't you think? Much more exhilarating to 'break away from.' I sense sometimes that the children see me as the 'failure' in the marriage. Often I see myself that way."

"I don't know," Trudl said. "The breaker-away can carry an awful load of guilt."

"Guilt, yes. I see that with Peter. But guilt can make you angry at the person you've broken away from too. Especially if it's been a unilateral decision, not a mutual agreement, that the marriage has broken down. Peter kept saying that he wanted me to understand that he didn't feel guilty, that in later years I'd thank him and so on and so forth, but I think he felt as guilty as hell. He said he should have gone years before, but I pointed out that he never would have gone unless he'd had someone to run to."

"Do you believe that?"

"Yes."

"But he's never lived with Anne-Marie."

"No. But she was there to give him hugs and cuddles. She told me that the night he left the island, the weekend we 'agreed' to break up, he came straight to her and said he'd just spent the saddest, most terrible day of his life. That's probably true, but it's awfully nice to go and be comforted afterward, isn't it? I remembered driving back up from the ferry to the house and knowing I was going to have to face the children. Hannah especially, because she was the only one who'd been told before he left, and face that empty bed and his old jacket on the hook, his pipe on the mantelpiece, his *presence*, which is so strong in this place and just wondering how I would manage or whether I shouldn't just take the kids and go away somewhere, quick, the next day. Somewhere where we could start again, I could start again."

"Why didn't you? I think that's what I would have done. Did do, really, by coming here."

"It's easier with one child. We were settled in. I hadn't enough money and would've had to borrow it, but who from? And where to go? And I guess — "

"What?"

"I guess I believed in my heart that he'd come back. I guess I believed he needed me."

Stella had been listening to all this, drinking tea and smiling, saying nothing. Now she said —

"If you had left completely then you might have got him back."

"Why?"

"You made it all very easy for him. Actually, you behaved like a mother."

"A *mother*!"

"Sure. Well, an ideal wife, anyway. You allowed him to do as he pleased, go where he would and come back anytime, all is forgiven. So civilized. So understanding. You wouldn't even take his money."

"I couldn't take money from a man who didn't love me. I was so emotionally dependent on him that I was afraid to be financially dependent on him too."

"That sounds good, but I think you were maybe trying to show him how wonderful and liberated you were. I've seen you leave a good soup or some homemade bread when you know he's coming over."

"I love him."

"You've got to be tougher if you want him back."

"She's right," Trudl said.

"Peter's not like that, I can't play games with him."

"You're just afraid to risk it."

"Maybe. Maybe the old alimony-and-accusation route was really more honest. Everyone says that our separation agreement is wonderful and how wonderful *I* am not to hit Peter for money and to share the children with him. But maybe I'm just trying to be Selene. Sometimes I'm so full of rage and despair I want to scream. I want some kind of awful revenge. But I'm ashamed of those feelings."

"Because you think you failed."

"Maybe because I know I did. But I'm often very bitter just the same."

Trudl put her hand on Alice's arm. "Somebody else is going to come along for you."

"I don't want somebody else, I want Peter."

"You idealize him, you know," Trudl said.

"Well," said Stella, putting out her cigarette, "he *is* pretty fantastic."

"Let's change the subject," Alice said. "I hear the kids coming back."

"It seems so strange to me," Alice said, to Stella and Trudl, "that you two went to high school together and are still friends. I have no idea what's happened to the girls I knew in high school. I guess I was never very close to any of them. I'm sure they all married young and had lots of kids. I can't remember any of them ever saying they wanted to be anything. Anything other than that."

"Did you?"

"No. Yes. I wanted to be a painter, funnily enough. But I

wasn't as good at it as I was at writing. I was a writer by default. But I wanted to be married first. Didn't want the one without the other. Can you imagine a man thinking that way? Can you imagine a man thinking, well, once I get married I can think about being a composer or a painter or whatever! Once I find the right woman."

"It's all changing."

"Is it?"

"Isn't it?"

"I think for most women the same order of priorities obtains. We just pretend it doesn't."

"Would you rather be with a man than writing your novel?"

"I want the whole works! I want to be with a man, not just any man but the miracle man, *and* be writing my novel. I want to be free to work six or eight hours a day and then play with my kids and then have supper miraculously appear and the children instantly fall asleep and the nannie look up from her knitting and say, 'Don't worry dears, stay out as long as you like,' and then my man and I go off down the road in the moonlight to make love in the woods."

"Oh *Alice*."

When Selene came back from her winter in New York she too had a guru and when she and Raven were reconciled and they came to visit the island each brought a highly colored guru photograph to meditate upon. They reminded Alice of the pictures she got as prizes in Sunday school. Selene and Raven stuck the photographs in the square panes in the new room, sat cross-legged on the floor and stared at their respective spiritual advisors. Alice thought the whole thing was crap; but because she loved her friends (and perhaps because she wasn't sure it was complete crap after all) she hesitated to say so until her friends had gone back home. Selene's father had died when she was three. Raven's father, for years, had been a dreadful alcoholic. When he was in one of his rages he would lock himself and his oldest son in the back bedroom, give Raven a loaded shotgun to hold across his knees in case Harry (Raven's father) tried to get out and hurt somebody before he sobered up. "Somebody" was Raven's mother, on whom, in his most drunken frenzies, he blamed all their bad luck. Raven sat there for hours, with the loaded shotgun on his lap. Sometimes he had to wrestle with his father; sometimes he had to knock him out. Now Raven never raised his voice, although Alice heard deep within it a kind of rumble, as of a suppressed rage which might someday erupt and spew forth something so terrible it would drown and cover whatever happened to be in its path. Selene, who loved him, never raised her voice either. At the age of sixteen she discovered she had had a false birthday for the past sixteen years, a doctor's checkup six months before she was "born." The first time Selene saw her mother after several years she lost her voice. Each letter brought on an asthma attack.

Because Selene didn't believe in adrenaline, if she had a bad attack she had to be nearly dying before she would consent to be taken to the hospital. Who was she trying to punish? What was she trying so hard not to say?

Alice's father. She could remember him pacing his room upstairs, the awful sound of his breathing. She lay in bed and knew he was going to drop dead any minute and that in some vague way it was all her fault. It was because she couldn't love him. She put the pillow over her head.

Selene was very angry when she found out her mother had falsified the date of her birth. She went everywhere, setting the record straight, to the church, to her elementary school, everywhere. Now she had forgiven her mother — after all, things were different then, you married a man and <u>then</u> you slept with him. They must have been very much in love. But still the asthma attacks continued.

"It is awful and awesome," Alice said, "to think of the power of our mothers have over us, no matter how old they are, no matter how old we are. The mother may have been reduced, through age, through time, to a tiny old lady who walks with sticks. But in our hearts, our psyches, she's still the giant shadow mother we saw reflected on the nursery wall. What is equally horrible to contemplate is that I might become, perhaps am already, that kind of figure to my children.''

''I don't believe that,'' Selene said. ''You're a wonderful mother.''

''Well, you can say that because you are an adult and also I'm not your mother. I know that I've made a lot of mistakes and will keep on making them. I know that they feel I'm responsible for the breakup of our marriage. They align themselves with Peter in very subtle ways and I try not to care. He has a certain glamor, just now. He's a romantic hero. It may be different next year, when they live with him. I don't know. There's a lot of that capital-M Mother in me. I fight against it but it's there. My doctor says all mothers are Jewish mothers, which isn't very nice but I know what he means.''

''I think your children are lovely. And that must have a lot to do with you, the way you are with them.''

Alice sighed. ''I'm going to miss them terribly when they're with him. But it's the right thing to do. He needs to live with them on a day-to-day basis too. Otherwise he becomes one of those sad, rather frantic men you see at Saturday movies, spoiling the kids and unable to really relate to them. Anxious. Irritable. Or overjolly.''

''What will you do?''

''I don't know. Cry a lot, probably. Maybe travel. Stella and I have a fantasy about Mexico. Going down there for a month or so. But Stella hasn't any money. We'll see. No doubt I'll be here most of the time.''

''Will you come and see us in our log cabin?''

Alice put her hand over Selene's. Raven and the girls were out on a walk to the end of the island. He had promised Flora she could ride on his shoulders if she got tired.

''I'd like that.''

Alice kept her hand where it was.

"You know that I slept with Raven while you were away."

"I know."

"Did he tell you?"

She laughed. "Do you remember when I came back, that first day, and you'd come into town with the girls? You were going back that night but when I arrived at the house you suggested I go back instead, have a quiet reunion with Raven."

"I remember." It had been a hard thing for Alice to do. Selene had arrived so refreshed and radiant after her months away. Alice was tired and discouraged. She had wanted to go back and spend a silly, carefree weekend with Raven. He had his canoe with him and had convinced her, finally, to go through the Pass with him, over to Coon Bay.

"I'm too scared," she had said.

"You won't be. Not with me navigating."

"Will you turn back if I am? Freaking out?"

"Of course."

(But she knew he wouldn't and she didn't care. Alice trusted him — so far as canoes were concerned — completely.)

"I got two rides up the island," Selene said. "And so it was quite late by the time I arrived here. Raven was asleep on Anne's bunk. I hadn't turned on any lights, just tiptoed in when I saw the house was dark." She turned her hand palm up, so that their hands were holding one another, palm to palm.

"I bent down and kissed him. He thought it was you. He said your name and held out his arms."

"Oh my."

Selene nodded.

"Then I realized how hard it must have been for you to send me here that night. But you had done it."

"Yes. I'm not quite sure what my motives were but Raven, much as I love him, is not the man for me. And I'm not the woman for him. You are."

"I hope so. I sure feel better about it now than I did six months ago."

"I wonder who *is* the man for me, if there is one — besides Peter, I mean."

"I don't think you were meant to be alone."

"I'm not sure. I don't think Peter's going to come back. He doesn't believe that I might have changed. And all the dope — it's like trying to talk to someone who is very far away."

"Or very near."

Alice withdrew her hand.

"That's where you and I differ, Selene. People who are stoned most of the time may be closer to *themselves* in some way, but I don't think they are closer to other people."

"I feel closer, I really do."

"You feel, physically, a heightened awareness. Of your body. Your surroundings. But I should tape-record all of us some time and play it back the next day. All the 'insights' and 'profundities.' Doesn't sound much different from drunks except the voices are softer. And you can't can't can't look after kids and be stoned. Or I can't. Somebody's got to be the parent."

"Why?"

"Selene, come and see me in a few years, after you've had a couple of kids. We'll have this conversation again, okay?"

(Five years later Alice, sitting at the kitchen in a rented apartment in Montreal where she was teaching for the winter term, opened a letter from Raven. A colored snapshot fell out. A picture of Raven and the second child, a little girl. Both were stark naked and smiling. In the letter Raven said that Alice was his fantasy woman. He thought of her and their times together when he masturbated. Alice had been feeling low. She missed the girls and the winter wind was howling around her ground-floor windows. She had had a lover and then she didn't have one anymore. But Raven's letter, and the photograph, made her laugh and laugh. Raven was still a child, he always would be. But he added that they were all snug and warm and happy and she believed him. She would go and see them when she got back to British Columbia. She wished he had sent a picture of Selene.)

MARCH

Trudl and Stella have gone to California. We are to take care of Trudl's cat and check on her place every day. Christobel is going to stay with Trudl's mother. They all left on the morning ferry. I felt very jealous at the ease with which they can arrange to take a holiday. (But be careful Alice, Peter would love to have the children *all the time*. Then you could take as many holidays as you liked; life could be one continuous holiday. Would you like that? You know you wouldn't. You'd be fretting like Lawrence's wife — "like a cat without her kittens.") I think I also felt jealous that they were going off together and leaving me behind. They never even said, "Gee it would be nice if you could come too." Probably because they know I couldn't but I'm feeling very much LEFT OUT and sulky. Stella is making this into a mythical journey, talks a lot about Isis and Osiris. Trudl (who has now definitely broken with Glenn) seems to be going along for the ride. *Her* California lover isn't dead, like Stella's, but he's an addict and she doesn't want to get mixed up with him again. They were all down here last night — Stella, Trudl, Harold, Glenn, Christobel, Uggah. Stella and Trudl took baths and washed their hair and then we had a bon-voyage party. They were both very excited and treated Harold and Glenn as though they were part of another world or another life. It occurred to me that they might not come back. Trudl could always send for Christobel or have her mother bring her down. Stella has no real ties, except to Glenn and her father and mother but nothing really to keep her here. She kept saying how good it would be to get off the island, see new faces and new sights. The party broke up early and I sat a long time by the fire after they all had gone. What will it be like now with them gone (for three weeks at least)? Who will I talk to, this diary? What if they don't come back? There's Glenn and Harold, of course, although I sense that Glenn may be moving on soon, but I need *girl friends*. I certainly felt, last night, that something had ended, but I may be exaggerating because I'm feeling jealous and left behind. I have never been on a trip with a girl friend, not since I was married. I know that this is a serious trip, not just a jaunt, but it has its jaunt aspects as well. I am feeling very sorry for myself tonight. Pick up one book after another, put them down. All these explorers searching for new routes, new lands, new markets for king or queen and country. I feel weighed down, anchored, not so much by responsibility as by my own inertia. I'm restless and yet the idea of starting new, someplace else, packing us all up — I don't have the energy. And besides, Peter and I promised one another we wouldn't do that.

(Not that promises can't be broken.)

INTO EACH LIFE A LITTLE RAIN MUST FALL

YOU HAVE TO LEARN TO TAKE THE
BITTER WITH THE SWEET

TIME HEALS ALL WOUNDS

LOOK FOR THE SILVER LINING

C'EST LA GUERRE

The mind a bird then, a glaucous-winged gull or some sea bird, flying low, skimming then quick, down. The silver thing, flapping, in its mouth. Memories. (But remember, also feeds on carrion and garbage. Follows the ferries for this very reason.) Drops molluscs on rocks to open their shells. Man on one of the other islands rushed to hospital with a great gash in his crown. Oyster fell on him, he said. And Henny-Penny with her acorn. Somewhere the sky is always falling.

The foolish man built his house
upon the sand
The foolish man built his house
upon the sand
The foolish man built his house
upon the sand
And the
Rain
Came
Tum
Bling
Down.

In the spring Alice had a business lunch with Peter in town. He mentioned that he'd been out for coffee with Anne-Marie. Alice concentrated on the books she'd just bought at Duthie's.

"Hmm. How was that?"

"Well, I think we were both a little sad. We're going separate ways now, but we still care deeply for one another. We felt we were like flotsam and jetsam which happened to be washed up on the same shore for a brief moment in time."

"Which were you?"

"Beg pardon?"

"Which were you, flotsam or jetsam? It makes a difference you know."

From his face she could see that he didn't think she was very funny.

MARCH

Not a word from Stella or from Trudl and two weeks have gone by. Glenn is gone to Nova Scotia, I'm not sure why; Harold is around but very much involved with the folks at Coon Bay. Uggah has disappeared. Now I am really finding out what it's like to be alone on this island with my three daughters. There isn't anyone else up here for me to talk to. Anne brings us gossip from the south end when she comes home from school. So-and-so's father yanked the phone out of the wall; so-and-so's sister is pregnant. There's a whole soap opera going on around us, equal to our "drama," I suppose, in triteness and intensity. An Indian girl got pregnant and shot herself in the stomach. She's alive and she hasn't lost the baby. How desperate she must have felt to commit such a desperate act. It goes way beyond abortion. Anne rattles it all off like the evening news. Everything gets equal weight.

When they go into town there is no more talk of Anne-Marie or mountain mix. No talk of Penny, either. I do not ask how they have spent the weekend (I know they see a lot of movies) and they do not volunteer any information. I no longer speak to Peter on the phone, it's too humiliating; I always cry. I'm supposed to be "over it" by now I guess and he gets annoyed. He has a set time that he calls to speak to the children and I let them answer the phone. Sometimes they say, "Dad says hello." Sometimes they say "Dad sends his love." I see him when I get off the ferry, after he's spent a weekend over here. He hands me the keys to the car and tells me where it is parked. I do not ask him to come back up the island with me; if he wanted to come he'd suggest it. He's probably too wrapped up in (or around) Penny to even contemplate such a thing.

I say "Hello there," trying to say it the way he used to on the phone, as though he's a neighbor I don't see very often. "Hello there." Then I hold out my hand for the keys and stare out at the water while he tells me where the car is. I go on along the wharf and up the hill, get in the car and drive home.

Tonight I felt like punching him in the face, really hard, breaking his glasses and then just walking away, leaving him to get on the ferry and get home as best he could. Perhaps the school-bus gossip stopped me. "Anne's mother punched Anne's dad right in the face down at the ferry dock. Right in the face!" I might need that but she doesn't.

And so I go quietly ("hello there") and come home and pick up a book. Lord Nelson again, being very English and brave.

"Now," said Lord Nelson, "we must trust to the great Disposer of all events, and the justice of our cause. I thank God for this great opportunity of doing my duty."

and then

"I'll give them such a dressing as they never had before."

Alice had a dream. She was in a large, old house. She had just put one of the children to bed in a new room and reassured her that everything was all right. (She was going to be sleeping quite far away, perhaps — she remembered upon waking that the rooms were far from one another.) There was lots of dark paneling. She went on down the corridor. In one room Peter was sound asleep, fully clothed, with the light left on. There was a photo album on the bed and photos on a table by the bed. She felt, she "knew," that he had been putting in a new picture that Anne had given him. She could see the old picture, blurred and out of focus, on a chair and thought how much better this new picture was. She was moved that he still cared about the "photo album."

Alice decided to tuck him up and take off his clothes. She did this very gently so as not to waken him, and yet she felt a tremendous desire to see him naked or perhaps just to "see his penis," as the phrase came back to her later. But when she slipped him out of his old trousers he had another pair on underneath ("very soft." She thought, in the dream, those words). They were Selene's old white trousers, which she had given to Hannah when she left. Alice knew this and was not surprised.

He woke up and they started to talk but she could not remember, later, what they had talked about.

The sheets were torn but patched with pretty patterns. She remembered thinking how nice they would look, hanging on the clothesline.

And if he had understood about her frozen heart? Or frozen shell, really, for her heart was a little fish swimming vainly to and fro, to and fro, beneath the ice. If he had come with a little tent, as they do, and a little saw, as they do, and made a fire to warm his hands (and a thermos of hot coffee in his rucksack) and sawed a circle and fished for it, that lost and frantic thing, swimming backwards and forwards beneath the ice of her despair.

But no. "She is cold," he said to Anne-Marie, to Selene, to Stella and Trudl, to whoever would listen to him. "She was destroyed before I ever met her."

Not understanding how fear and pain and all those vast quantities of blood, and the little dead thing in the basin had frozen her over, had tossed her out beyond her depth, where only the sea gulls mocked her cry of "me me me." So that when she returned to life, was flung back in, she had suffered a sea change of a most terrible sort and was dumb to tell about it.

He would move away from her in the bed, sleepy. "Alice, don't put your cold feet against me. Please." And she would lie sleepless at the very edge of the bed, icicles forming along her frozen cheeks.

MARCH

I had to kill a litter of Tabby's kittens today. All but one which I managed to give to Anne's friend Maggie. Glenn once said he'd heard it was much more humane to put them in a sack and tie them to the end of the car's exhaust pipe, that they would die in a couple of seconds, a far better method than drowning. I waited until the girls went into town. I waited until it began to get dark. Then I shut Tabby up in the house, ditto Byron, and got a burlap sack from the back porch. Nobody said they were willing to take them when they were old enough. Nobody wants another cat. People come over here and leave their cats to fend for themselves—or to die—as it is. Peter won't even have Tabby. He says two of the kids who are going to be in his new commune are allergic. (He won't have Byron either, pointed out how I'd always said the city was no place for dogs.)

I sat in the car, running the engine, for about three minutes. Just to make sure. But even so they weren't dead—maybe the burlap was too porous, I don't know — I could hear a tiny mew-mew—like a bird, or birds. I wanted to run away. I wanted to scream and scream. But I took the sack down to the boat launch and without looking into it, I thrust in stones and sand. Then I tied the neck tight and walked nearly to the end of the island, where I knew the current would be terrible and swift, and I threw it out into the water.

Now it is 3:00 A.M. There is no moon. Tabby prowls the house (I burned the kittens' box and blankets). I dare not sleep. I dare not even close my eyes. I want to be held and comforted; I want to be told I did the right thing, the only thing. The window is open. I can smell spring out there. It sickens me. The whole world sickens me. I want to call up Peter; I want to tell him I really can't go on. Everything affects me so. I'm afraid.

When Stella and Trudl came back from California they brought Alice a book of essays on problems in perception. They each had new handmade sandals and new jerseys that laced up the front. They were both smiling even though they'd had a rotten time. The pear tree was in bloom and the air was full of bees.

In one of the essays Alice later came across this phrase: "If you are sane you know that the word 'cat' cannot scratch you."

Peter and Stella and Trudl had decided to paint a mandala. The original idea had been to give Peter only one twenty-eighth of the canvas because he only came to the island once a month. Later, as it became more serious, they each decided to take a third. Stella showed Alice a drawing she had made in her journal: Peter in the middle with his arms around Stella and Trudl.

"Where do I fit in?" said Alice.

Stella didn't answer. "It's only a drawing," she said.

Stella and Trudl began going into town once a week to meet with Peter. Sometimes they went to a restaurant; sometimes they went to Alice's house.

"I feel really funny about all this," Trudl said on the

phone one day. Alice had invited her and Christobel to come up for supper the next night but there was a mandala meeting and she couldn't make it.

''I feel really funny about it too. I feel *very* funny about it actually.''

''It's just become so important. The most amazing things are coming out. It's scary. Well, you understand that, don't you? The same thing happens to you in your writing.''

''I don't write with two other people.''

''It's different. You've been doing it for years. We're just learning. Even Peter says that about himself.''

''Trudl, maybe I'm just paranoid, but I don't like it. I don't like what's happening to you and me and Stella. I hardly ever see either of you any more and when I do, you act funny.''

''Alice, you were the one who put the sign on your door last fall. I always admired that — your dedication and self-discipline. I sometimes wanted to see you but I respected your privacy.''

''It's not the same thing.''

''I think it is.''

''Well you're wrong.''

That night Alice decided that she would have to speak to Peter.

Alice put a notice above the telephone.

''There's no such thing as a free lunch. All long-distance phone calls must be paid for the day they are made. Get time and charges from the operator. And don't come in my fucking house when I'm not home!''

''Some of the people at Coon Bay were really upset by your sign,'' Trudl said.

''Oh really?''

"You could have worded it better."

"I worded it the way I wanted to word it. Twenty dollars in calls that 'nobody' made."

"I guess you worded it all right."

"I guess I did."

"They're not all bad."

"Of course not. Most of them are fine. But that won't get me back my twenty dollars. They think that because I have running water it's okay to rip me off."

"Well, you've made some enemies."

"Trudl, I am not Selene *or* you. It doesn't worry me to have a few enemies. You like them so much, you get a telephone."

"I guess you're right."

One Sunday, when Alice came back on the morning ferry, Peter was still there, sitting on the front porch drawing. His rucksack was beside him.

"Oh, hello there."

"Peter, you didn't have to wait. I told you I'd be back this morning."

"Yes, well. I have to go to a meeting at the south end around one so I stayed for that."

"A meeting?"

"About the mandala, actually. Trudl invited Stella and me to lunch and then we're going to discuss the mandala."

"So you'll be on the island until tonight."

"Yes. But I might as well stay down once I'm there. You can just carry on, don't let me disturb you. I caught some cod. There's quite a bit in the fridge."

"Thanks."

He sat there for two hours, ignoring Alice, who offered him a coffee but he said no thanks he'd had one just before she arrived. Alice took Flora to the boat launch to play in the tide pools. She heard Stella's car go by and, a few minutes later, honk down below their path.

"Let's see if we can find another hermit crab," she said to her youngest daughter.

"All women are lunatics through biological determinism," Stella said. "I wonder if, originally, all women bled at the same time, and in perfect harmony with the waxing and waning of the moon."

Alice laughed. "It's useless to speculate on that, don't you think? But it would be interesting to know if women whose cycle *did* correspond to the moon's, now, were more stable, or less."

"Yes!" Stella said. "We could send out a questionnaire, they wouldn't have to sign it or anything."

Trudl shook her head. "The trouble is, a woman's cycle might correspond at one time and not at another. If she's had an abortion, or a miscarriage, or a child, the whole thing can change drastically."

"It is weird, isn't it?" Stella said, "the twenty-eight-day thing."

"Well, only the moon is *that* regular!"

Alice poured more tea. "You know, the tides have a point where there's no perceptible rise or fall, a kind of still moment. There must be that in women too, a moment, maybe an hour or a day, when ovulation is just about to begin, but hasn't. That might be our most stable time if we could locate it precisely."

"How could we?" Stella said. "I can tell when I'm getting my period and I can sometimes tell when I've ovulated, I get a pain — "

"Yes," Alice said. "There's a German name for it — *mittel schmerz* — 'middle sorrow.'"

"Is that true? My parents are German," Trudl said, "and I never heard my mother use that word."

"I read it somewhere. But we might be able to find the point just before that, the same way they can tell when slack tide, that still moment, begins. The *mittel* without the *schmerz*."

"It's probably been done," Stella said. "Using thermometers."

"I can't see it would really benefit us," Trudl said.

Alice shrugged. "Maybe not, but most women seem to be very affected, emotionally, by their periods. If there was a brief time of great stability, we could put off certain decisions until that time. Or even certain arguments."

"I get terribly depressed just before I get my period," Trudl said.

Alice agreed. "So do I. Horribly. Very moody."

"I get very horny when I've got it," Stella said.

Alice laughed. "You would! But actually I think a lot of women feel that too. Probably because they know they can't get pregnant."

"But you can't get pregnant when you're on the pill anyway," Trudl said.

"True."

"But it's a bitch, isn't it? All this bleeding and Tampax and pregnancy stuff."

"Our sister moon doesn't have to go through any of that!"

"No. It's the sun that bleeds, isn't it, across the sky. Bursts and bleeds."

"And the moon is never in heat. Chaste Diana."

"The frigid moon."

"It must be awful to be frigid."

"I doubt if that's a problem you'll ever have to face."

"No. I've got the opposite problem."

"During the really bad days," Alice said, "I'd open the door to go for a walk and I'd be slammed back."

"Slammed back?" Stella said.

"Yes. Almost as though there was a big wind out there pushing against me. Like coming around a corner on the ferry sometimes. Whenever I'd try to do something independent I'd be slammed back."

"Who was slamming you?"

"Life."

"I was walking along a street once on acid," Stella said, "and I saw the name 'Edge' on somebody's front lawn. I didn't dare go any farther."

"That's it. You were slammed back *then*."

"When I went to get on the plane last year to go to Berkeley and couldn't get on, I guess I was slammed back then."

"I wonder why you were able to go this year?"

"Well, Trudl went with me. Or at least got me down there."

(Stella wrote to Alice — "I sit in cafés waiting for pieces of Robert to walk in.")

When Stella and Harold and Trudl and Christobel dropped in Alice was just going to serve the curry.

"Why don't you all stay? There's plenty."

"I don't think so," Stella said, smiling. "As you know, I hate to wash up." But they all stayed anyway, sitting on the floor in the new room.

"We found where we're going to do the mandala," Stella said, "we just called Peter from dad's."

"Oh, where?"

Stella looked at Trudl. Trudl looked at Stella. They smiled at one another.

"We can't tell," Trudl said.

"It's a secret."

Alice was seven years old and in the girl's cloakroom. "Two's company, three's a crowd."

"I'll see if there's any more rice."

Trudl was showing the last of the NFB films so they walked over together to the community hall. Then Alice realized she had forgotten her glasses and had to go back. She met Stella on the road.

''Every last dish done.''

''Thanks Stella. I know how you hate it.''

''Peter called.''

''He did?'' Alice was horrified at the birdlike quality of her heart. It wasn't his day to call.

''Yes.'' She hesitated. ''He really wanted to leave a message about our mext meeting. It was lucky I was there. Save you the trouble of walking up. He said hi.''

''I see. Maybe you'd better tell him to leave his messages somewhere else.''

''Oh *Alice*.'' Stella put her hand out helplessly.

''Excuse me,'' Alice said, ''I have to hurry or I'll miss the film.''

The trouble was, the Mandala Club was to be an exclusive and secret society of three.

The trouble was, two of the three were Alice's best and virtually only friends on the island.

The trouble was, the third was Alice's husband.

She dreamed of five enormous blue-red giants with a single eye.

She dreamed that someone was crying bitterly in another room.

She dreamed that Peter's face was between her thighs.

She woke up from an unremembered nightmare so shaken that she decided to wrap a shawl around herself and go make a cup of tea. She was snuggled in bed drinking her tea, the covers up to her neck and beginning to relax when she looked up and there were two veiled figures at the end of the bed, one male and one female and they were talking rapidly, in DEAF, about her.

She imagined, she did not dream, that one of the hunters up on the ridge had come down the water mains and was greedily staring in at her while she slept. When she looked out he was there in his red-plaid shirt, underneath the lurid moon. He broke a pane of glass with his fist and she woke up moaning with Byron anxiously licking her hand.

In Berkeley Stella fucked three men in one afternoon, looking for pieces of her dead lover.

"And?"

"Well, I'm here, aren't I?"

"It didn't work."

("Sitting in cafés, waiting for pieces of Robert to walk in.")

All the broom came out along the roadside. Alice and Trudl discovered one day how the weight of the bee landing on the flower opened it up like lips, and exposed its nectar.

"In *Women in Love* there's a wonderful scene where Birkin describes flowers to Ursula's schoolchildren," Alice said. "Perhaps cocks and cunts are flower symbols and not the other way around." They stood and watched the bees.

There were purple grape hyacinths in the grass and paper narcissus and daffodils. When Alice was in town she bought some chocolate eggs to put in the hen house Easter morning. She would have to keep them wrapped or the hens would shit on them. The foil could be removed.

"Reality versus romance," she said to Stella.

Stella and Harold had walked over for a visit. Stella was cross because Harold had been decorating her with flowers.

She kept shaking her head and releasing a shower of petals.

"How's the novel coming?"

"Terrific," Alice said, lying. The three women were restless. They looked at one another with hooded eyes.

It was all this blooming. The very ground was juicy with it, underneath their feet. They had strange dreams which they told to one another anxiously. They began to try and see if they could have the same dream by meeting last thing at night and deciding on a subject.

"But there's only one subject, isn't there?" Stella said.

The preliminary work on the mandala took a lot of time. Alice hardly saw Trudl (who was staying at the south end now) or Stella any more. She worked in her garden and talked with her daughters and told herself it really didn't matter.

"Are we going to plant pumpkins again this year mommy?"

"Of course. Maybe some white mice too."

"White *mice*?"

"I was just teasing. Remember Cinderella? Did you know her famous slipper was really made of fur? The translator made a mistake, or so I was told."

"It's not nearly as nice with fur. Sounds like a bedroom slipper."

"No it isn't, is it?" In the distance she could hear Stella's rattly old car coming down the road on its way to the south end.

"Come here and help me weed these lettuces," she said, "then I'll help you plant your pumpkins."

The car went by the path without stopping.

If only I could learn to hate him, Alice thought to herself miserably. If only I could. Digging furiously with her little pointed spade.

Alice drove down early to the south end one morning to talk to Trudl. She answered the door with just her panties on, her long red hair disheveled.

"I have to talk to you Trudl."

And Trudl weeping as they sat at the kitchen table,

"He left when I was thirteen. I watched my mother's pain for years. Now I watch you. I can't go through all that again. You open up too many wounds. I was sure it was all my fault."

"Are you sure that's it?"

"That's an awful lot to do with it; it isn't just the mandala."

They hugged at the door and Alice drove back up to the cabin, thinking.

Remembering, as a child, those little bloody dots in eggs.

"What's that?"

Her mother, always reluctant to discuss sex, said,

"It means that a chicken has started."

"What do you mean, 'started.'"

So Alice agreed to no rooster, just the four white hens who were all named Henrietta.

Then in the late spring Glenn came back to the island and the next day arrived at the cabin door with a large cardboard box.

"We'd better open it outside," he said anxiously. Henry went in the chicken run with the others. Alice and the children threw in handfuls of vetch and vegetable peelings. In ten minutes it was clear who was the boss. All the Henriettas ran around dizzy with adoration, a collective bird frenzy of waiting on and preening.

"Look what you've done, Glenn," Alice said.

Henry gave a crow that was almost a yodel.

"Why don't the hens crow like that?" Hannah said.

"I don't know. Strange hey? All dogs bark, all cats meow, all canaries sing — I think. We'll have to find out."

Henry would begin rehearsing for dawn in the middle of the night. He was not popular with other people but the

Henriettas were very productive and proud.

Alice remembered one of her African sayings:

''The hen too knows that it is dawn
But she leaves it to the cock to announce it.''

Although there were posters everywhere, Alice did not see signs of anyone marching to a different drummer, or not in the solitary sense that Thoreau meant. When she was in the city she bought *Walden* and *Civil Disobedience* and gave them to Hannah to read.

''Of course,'' Alice said, ''he didn't stay there very long and rumor has it that his aunties paid his taxes. But he did *observe*, he did think about things and try to organize those thoughts, to share them with others. I'm not saying one has to actually sail the seas or climb mountains to make discoveries, but there's something so passive about what's going on now. I was trying to explain it to Raven one day. There's no real spirit of adventure; there's no reflection either.''

Hannah looked up. ''I don't know what you mean by 'reflection.''' Hannah had become a vegetarian. Raven and Selene were her ideals.

''Look. If everything is 'far out' or 'oh wow' then somehow it's all trivialized. When you are stoned most of the time it's as though you lie in this warm bath of benevolence, or maybe even in the warm bath of the womb. You lose all your energy. Yes, it is like staying in a warm bath too long. It isn't just your fingers that get wrinkled. Your mind gets wrinkled too. You sit around and sit around and nothing, really, in spite of the oh wows and far outs really moves you to the core. You're too 'mellow' for that. Raven and Selene, who by the way, have a lot more intelligence than most of these people, talk a lot about 'nonattachment' but they are in fact very attached to a lot of things, including one another. But they've been sucked in—they want to be children again, free, playing in meadows.''

"What's wrong with that?"

Alice sighed. "I don't know what's wrong with it. Maybe I'm just jealous. Maybe I'm just an old fogy. But I don't think so. There's something very bland and colorless about all this — what they say, what they do. They're flabby, somehow." She held up her hand, anticipating Hannah's objection before she even opened her mouth. "I don't mean they aren't beautiful and don't have beautiful *bodies*, most of them. I mean there's a flabbiness of spirit. Any minute now I think their minds are going to sputter and go out."

Hannah shrugged. She had always been quiet but since moving to the island, since Peter's departure, she was even more so. She and Peter smoked up together, had little grown-up talks about 'love' and 'caring.' Alice was worried about her but didn't know what to say. She knew Hannah thought she was cynical. She knew Hannah disapproved of her sleeping with Raven, even though it had only been occasionally. Sometimes she thought she heard Hannah crying in the night, but when she got up and went in to her, whispered, 'Hannah, are you awake, do you want to come and sleep in my bed for a while?' there was never any answer. Alice had a friend whose child became a diabetic at the age of ten. He told her he often thought he heard the little boy sobbing only to get up and check and the child was sound asleep and peaceful. The child was dealing with it; he was not. "Children are resilient," he wrote, in reply to an anxious letter from her. "They deal with things in their own way, which may not be ours. Perhaps you *want* her to be wide-awake, as you are. Perhaps you want her to take some of the burden of your grief. She may, but don't insist upon it. That's like a parent saying to a kid who's learning to ride a bike, 'Watch out, don't fall!' Pamper yourself a little — the kids will be okay."

Was that true? Sometimes, with Hannah doing her schoolwork at one end of the kitchen table, Alice at the other, writing, there was a kind of peace. They took turns putting wood in the small cook stove, stirring the soup.

They figured out a vegetarian substitute for the meat dish the home-economics course wanted, argued their case and won. They put crumbs and sunflower seeds on the bird-feeding station outside the window and looked up, from time to time, to watch the chickadees and juncos, the three raucous crows who moved in like a black-leather motorcycle gang and drove all the little birds away. They decided to write a fairy tale for Flora. Alice would actually write it and Hannah would illustrate it. The crows were really princes under a spell.

Alice went down the path and across the road to pick up Flora. Flora was developing a passion for Spaghetti-Os out of a can.

''I think she has them every day for lunch,'' Alice said. ''And Cheeze-Whiz sandwiches on white bread. But I can't really send her down with a packed lunch; it would be too rude. I hope what we eat cancels all that out.'' Hannah wasn't sure, but she didn't get up on her high horse about it, just teased Flora in a nice big-sister way and took her for a walk or in the front room to play while Alice turned the ''study'' back into a kitchen again.

At three-thirty the yellow school bus stopped at the store and Anne got out, sometimes with her new best friend Velma. Velma lived on the reserve and was as dark as Anne was blond. They had a snack and went outside to do the kindling.

''Velma thinks you're nice,'' Anne said one day.

''Good,'' Alice said. ''I need some approval.''

''But her mother doesn't understand why you want to run around with the freaks from Coon Bay.''

''Her mother told you that?''

''No. Velma told me.''

''Well tell Velma to tell her mother that I don't run around with the freaks from Coon Bay. I'm curious about them and sometimes, on mail days, I invite them up for tea. But I don't 'run around' with them.''

Anne looked uncomfortable. ''Okay, okay, don't get mad. I'm just telling you what Velma told *me*.''

Alice laughed. "Okay yourself. But sometimes it seems to me that it's heads I lose, tails I lose. Anyway, let's have supper."

Flora pulled at her mother's sleeve.

"Can we have Spaghetti-Os?"

Alice looked skyward.

"Give me strength," she said. "If you're up there, give me strength."

APRIL

"It is a strange thing, being deaf. Some people call it the invisible disability." I read that in a book. Harold, coming up the path with Stella and Trudl, looks just like any other large ugly-attractive young man. Indeed, what you would notice about him first is his mop of curly red hair. He has the kind of hair to which the word "mop" really applies. You might also wish he had had braces on his teeth when he was younger. But when he begins to speak there is something not quite right — "Mom, is that guy stoned?" No, or not at the minute, he is simply deaf as a post, as a doornail — as a stone, if you like. Raven's missing finger is more obvious than Harold's deafness and yet it doesn't bother him at all.

Most of the time Harold is cheerful, accepting. He says to me, Stella, Trudl, the girls that we are the unfortunate ones because we have to "listen to all that shit." He tells us how he used to ride the buses in Vancouver with a friend of his. They would pretend they were mentally retarded; it was easy to do because of their voices, because their real handicap was invisible. They would pick their noses, drool.

"Wunnerful," he says, grinning his big ugly grin.

But I have seen that he does not always know where to look in a group discussion. I know that in spite of his almost uncanny ability to read lips, whenever we turn our back on him, he will be deaf again. (Do the deaf speak to you in dreams? Do their lips move? Can you read them? I don't dare ask.)

Stella complains that the deaf are very noisy. She is getting somewhat fed up with her role of companion to a deaf man. And their sex life, she says, is pretty ho hum. Her words. Stella has been with a lot of men so she should know. She is getting restless. She and Trudl are thinking of taking another trip. When Peter spends his weekend on the island, I hear of long suppers where wonderful Peter listens to them and doesn't laugh. I want to tell them, "I don't want to hear about it," but I do, I do. What he cooked, what he said, how wonderful he is. It helps to feed my pain, which I now see as a dog with an enormous bright red mouth, a mouth like a fiery furnace, demanding to be stoked.

Sometimes Peter makes love to me in dreams.

Harold and Peter get along famously. Harold can always make Peter laugh. "I really like those two," Peter says. "They're quite crazy." Stella has begun to flirt with Peter. Very subtly but it's there. I can see it, can sense it, and it hurts.

He had a lovely thick cock, solid and dependable as a walking stick. But didn't like his body, no. Also confused somewhat about mothers and lovers. "It's easier for us," Alice said to Stella, "nothing comes out of them but come. Babies drop out of us from our most secret places, through channels the fathers have charted and laid claim to. Rivers of pain and blood. Would they do that to their own *mother*? Forbidden. Not nice. Never see mother except above the waist.

"Ma-ma, the breast. It's the same in all languages. Babies with mouths like flowers, turning toward the sun. Rivers of white light and then, sated, satisfied, the small head falling away, the mouth slack.

''Men are related to the sun. The sun never changes his shape. Sisters of the moon we are, shape shifters but oh so predictable in our shifting. We hold the waters of the world in our nets.

''Once I became a mother — actually once I became *pregnant,* he had trouble. And yet he is torn, is at heart a family man. Do all men want both — a mother for their children, a wife and a mistress?''

''Wouldn't you like that too?'' Stella said. ''Come on now, wouldn't you?''

''I don't know. Anyway, I don't have either, do I? The question is academic —or is for me. I don't think I wanted anybody else. I just wanted things to get better.''

Three beautiful rosebushes grew by the side of an old white shed, Peace roses, they were called. Yellow with pink splashes, they reminded her of certain delicate sunsets, or of peaches if they were suddenly to bloom.

They started in June and they went on and on, sometimes until late September. They had huge thorns, like cats' claws, and a cool rich smell. Alice filled mason jars and wine bottles with them, gave them away in huge bouquets. Her father had grown roses —her grandfather as well—but she knew nothing about their care. They were planted by the old man who used to live in the cabin and who now lay in the small cemetery at the south end, the cemetery where Alice often thought she would like to be buried. The cemetery where she went in spring to pick morels. Morels grew well where bones lay underneath.

Every year Alice cut the roses back and watered them carefully, snipped off the dead heads and planted chives as protection against the aphids. And every year they gave her back far more than she gave them. They seemed to bloom for the sheer voluptuous pleasure of the blooming, the bursting forth.

Rose again from the dead, he. First entry under ''rose''

in her *Bartletts Familiar Quotations*. She had never made that connection before but was looking up the word to find out something else. Rose of Sharon, was it?

"He descended into Hell. The third day he rose again from the Dead: He ascended into heaven. And sitteth on the right hand of God the Father Almighty—"

The child Alice in the choir of the First Presbyterian church. Waiting impatiently to sing "Love Divine, All Loves Excelling." The choirmaster had a wooden leg and Alice was always afraid she would be asked to sit on his knee and would feel his device through her skirt. Or bang it with her patent-leather Mary Janes. But couldn't stay away from choir because of the music, because of making a joyful noise.

Not much was made of Mary in the First Presbyterian church. That was for the Catholics and the kids who went to Saint Pat's. These nice people in their muted clothes, their hats and gloves, probably didn't give her a thought. Except at the Christmas pageant which went along with the annual white-gift service. Then there she was, represented by the prettiest girl in the primary division of the Sunday school. The chosen girl had her hair put up in rags or bread papers the night before. And wore maybe just a touch, the merest hint, of rouge?

Dear Mary, she looked so well after her confinement. Nothing blood-stained or weary about her. "Sweetly pretty," as Peter's mother would have said. A rose without a thorn. (Without a prick at any rate.)

Alice's father sang songs with strange lyrics:

"Oh the Bulldog on the bank
And the Bullfrog in the Pond
If the Bulldog met the Bullfrog—"

Or,

"The only girl I ever loved
Had a face like a horse and bug-gy

Be careful of that monkey wrench
Your father was a nut.''

''Alice,'' he said, ''I can row a boat, canoe?''

He had grown roses, never red ones, just soft-colored ones, pinks and yellows and peaches. When she visited her mother six weeks after he died, there was a bunch of his roses, overblown, in a water jug on the dining-room table.

Her mother told her that his last words, frightened and in great pain, in an oxygen tent, were

''Fuck you.''

Alice sat there, not knowing what to say. Her mother was very angry. It must have seemed like a curse. Alice was supposed to lift it.

They were sitting at the dining-room table drinking Nestle's Iced Tea with Lemon Flavor. Alice had not been able to come to the funeral; she had not yet seen the grave.

''Do you remember the year I ate rose petals? I must have been six or seven. I can't remember now whether I really liked the taste of them or whether I was being a smart ass.''

He had been given a grand Masonic funeral. When Alice left, her mother gave her his Knights Templar watch fob; the facepiece on the helmet really moved and the compass and mystical letters, which had always fascinated Alice, were still a mystery.

K. S. H. T. W. S. S. T.

Whenever she asked her father he always said it meant: King Solomon Had (a) Thousand Wives, Some Say Twenty.

Now, would she ever know?

Alice stood by the front window, waiting for the taxi. She had grown up in this house, or at least grown taller and older. Her father had lived here as a child. He must have loved it; but she had always felt oppressed by it and was glad it was going to be sold. She knew she should try and say something to comfort her mother — "He didn't mean it," "He loved you," "Remember all the good things." The trouble was, he probably did mean it. The more she had thought about it, the more shocking it seemed. It really *was* a curse. Her father going into that final darkness and having one last tantrum.

"No end to the withering of withered flowers."

As the cab drove down her old street and turned in the direction of the airport, Alice had a thought which almost made her ask the driver to turn around. But she would miss her plane. Peter and the children were thousands of miles away; she wanted to go home.

"What if," she thought, "he wasn't speaking to mother at all, but to God?"

"I didn't treat him with respect," Alice said. "Married couples forget all about their party manners. I don't mean pretending but just the extra little hugs and kindnesses because you sense the other person needs it. I wasn't — I wasn't *tender*. We were very comfortable together, a good team. But I never let him know that I found him exciting."

"Did you?" Stella asked.

"Oh yes. In every way. But he seemed so closed off, often. So private. He made me feel inferior because I yelled or was moody or unfair. He wouldn't fight, ever. And of course *I* saw him as superior, so much more 'civilized' than I was. I always felt he was judging me, that I didn't live up to the ideal housewife or ideal mother or ideal lover. Now I wonder if a lot of that wasn't my own insecurity, if I wasn't judging myself. Wanting to be all those things and feeling I'd failed. I could never seem to keep the house tidy or the refrigerator cleaned out or the kids' braids even or make anything with my hands. I guess I wanted to be good at all those things because his mother was. Or because my own mother wasn't. What I didn't see was that Peter loved me, if he did, for all the ways in which I was different from his mother. He hates his mother, hates both his parents I think — their small-mindedness, their excessive cleanliness and emphasis on appearances.

"Of course, now that he's been born again he *is* judgmental. Doesn't hesitate to feel superior, capable of deeper and finer emotions etcetera. Wants to be with people who don't lay any trips on him. Selene is really his ideal. I think he's found out about Anne-Marie."

"But Selene is pretty remarkable, isn't she?"

"She's a beautiful person. She also is a natural-born martyr. Her relationship with Raven is an awfully strange one. He calls all the shots. And look at her terrible asthma attacks. She never raises her voice, never gets angry, and yet she suffers from severe asthma. I think there's a lot of anger in Selene that she's afraid to let out. She too is trying to live up to some impossible standard she's set for herself."

"Maybe all women do it," Trudl said. "I know I feel terribly guilty about my anger and mean thoughts. I was brought up to be 'nice.'"

"I never feel I have to live up to any perfect-wife shit," Stella said.

"You've never been a wife."

"True. But I don't mind being lazy or mind admitting that I hate housework and cooking and sewing and never want children."

"You are extraordinary, you really are."

"I'm beginning to think so."

"Maybe you always had so many men around that you could afford to be yourself."

"I always saw myself as a mistress, never as a wife."

"Not even when you were a teenager?"

"I don't think so. Never went steady or any of that stuff. Perhaps my parents' marriage put me off."

'Well why didn't *my* parents' marriage put me off?"

"Or mine?"

"I wonder what our breakup will do to my kids? It makes me scared, sometimes."

"I don't know. I think you've handled things pretty well."

"Yes. Very civilized we are. The show must go on."

There was a short silence.

"Let's have a tea party tomorrow," Trudl said. "We can have it at my place. Flora and Christobel can bring all their dolls."

"We can dress up."

"Let's send invitations to the kids."

"Let's not invite any men."

"Sometimes," Alice said, "I'm not sure we aren't going right back to the beginning."

"What do you mean?"

"Well, are we ladies or are we only playing ladies?"

"Well it's more fun to play one than be one."

"I'll wear my Spanish shawl," Stella said. "And maybe read tea leaves."

"I'm going to bake a chocolate cake with chocolate icing."

"I'm going to curl my hair."

"Hannah said you were in your playpen," Stella said. She came in and shut the gate behind her.

"What are you planting?"

"More lettuces, a different kind." Alice began to curse as she moved along the row.

"What are you doing?"

"Cursing the seed—it's an old custom Selene wrote me about and I thought I'd try it."

"It gets really complicated doesn't it, all this organic shit and planting by the moon."

"I guess it's been done for centuries but of course its fashionable now—or with the counterculture at least. I know that my friends in town have been planting by the moon."

"Did you buy organic seeds?"

"I bought a few at the Golden Bough. But they're terribly expensive. The rest I bought all over the place, whenever

I was in town and thought about it. Did you?"

"No, I just took whatever dad had left over from his garden. It's fun isn't it? Working in a garden."

"Yes. Selene says plants are easy because they let you love them as much as you want without getting all emotional about it."

"That's nice. She's really an interesting person isn't she?"

"We all are. You, me, Trudl, Selene. What I want to know is, where are all the interesting men, now that all these interesting women are ripening like fruit?"

"They must be somewhere. This island isn't a very good place to meet men."

"So we keep saying. But a good place in every other way. I agree with you though. It's a bit as though Adam had been banished from the garden and Eve's punishment had been to stay there all alone."

"With the snake."

"Yes."

"You've got lots of things coming up already."

"You know, I thought a lot about how it was all going to look. A kind of enormous still life. Only not still at all. Things growing and swelling, changing color. I've planted nasturtiums all along the back fence. And gourds in the corners. I told you I always wanted to be a painter. Of course, being me and in spite of using string, none of the rows are really parallel."

"Ilma Hayes puts a starfish under each of her tomato plants," Stella said.

Alice nodded. "I've heard about that but somehow I can't do it. I know there are too many but I can't just bop them into the ground. Bury them alive."

"You could kill them first."

"How do you kill a starfish without just sticking it in the sun and letting it rot. And then they stink so!"

"The purple ones are really beautiful," Stella said. "I'd never seen them that color until I came here."

"They're very strange creatures," Alice said. "Grow back their legs — legs? arms? *points* — if they are lopped off. Since we came here and I've been taking Flora down to the rocks to play in the tide pools I've become fascinated by all that stuff. The little crabs, and the hermit crabs, and tiny fish. Even the barnacles. I never knew they were alive before. And all the different kinds of seaweed. I've used a lot of seaweed on the garden, a Raven-and-Selene tip.''

''Did you know we all have gardens now? Dad gave Trudl and Christobel a corner of his.''

''That was nice of him. I can't help liking your father.''

''Everybody likes him; he's a lovely, funny man. Everybody except Glenn. Glenn can't forgive him his drinking.''

''Is that why your parents split up? Because of the drink?''

''Oh. It's a chicken-egg thing. I don't know whether he didn't start drinking because they were unhappy together.''

''But you said he's been like this for years.''

''They just waited to split up until we were all out of school.''

''It's all pretty depressing.''

''Marriage seems to be a fatal mistake.''

''I still believe in it. Peter can say all the shitty things he likes but I think we had a good marriage. It's just that what he wanted was a good affair.''

''I used to tell people you and Peter were the perfect married couple.''

''As a compliment you mean.''

''Absolutely.''

Alice wiped her hands on her trousers and picked up the string and scissors. ''Let's go have a cup of tea.'' They went out of the garden and Alice shut the gate carefully. Alice looked at Stella.

''I'm going to put up a sign this weekend.''

''Saying what?''

''From *Under the Volcano*:

> Le Gusta Este Jardin:
> Que Es Suyo?
> EVITE Que Sus Hijos Lo DESTRUYAN!

You like this garden?
Why, is it yours?

We evict those
Who would destroy.''

''That's pretty strong,'' Stella said.

''I'm just tired of people coming around and helping themselves to things. Flowers — the telephone. Maybe it is strong. Maybe I'm just feeling grumpy. Peter probably won't keep up the garden anyway. He'll be too busy with the mandala.''

Stella pulled up some grass at the edge of the garden. She avoided Alice's eyes.

I spy with my little eye, Alice thought. But she too said nothing. Or nothing about Stella and Peter and Trudl. Instead she said, ''Hang on a minute, I have to put the hens away. There's a raccoon that sits up in a tree on that back road just waiting for me to make a mistake.''

Stella waited, chewing on a long piece of grass.

''I'm smoking again,'' she said.

''Oh *Stella*.''

''That's one of the reasons I came over. Besides to say hello. Can I roll a couple?''

''I'm going to get rid of that Player's tin.''

''Trudl's smoking again too.''

''You're both terrible. Come on.''

''She had a nice phrase the other day.''

''?''

'''Lately I feel I'm at the bottom of the lake.'''

''I guess I've been living down there so long,'' Alice said, ''that I don't even notice it any more. No, that's not true.''

Just before they went into the house Alice said,

''You know, I even pray for Peter to come back to me.''

She was immediately embarrassed to have admitted it and, being in a cursing mood, cursed her eyes, which had filled up with tears.

The toilet stopped working. When Alice was in town she bought a Jiffy Home Plumber but that didn't do any good either. For a month they used the outhouse at the community hall and peed on the ground. She got to like getting up, if it was a clear night, and going outside to pee beneath the stars. But when it was rainy and cold it wasn't so much fun. Stella found a flower-decorated chamber pot at the Goodwill in Victoria and donated it to the cabin but Alice knew something would have to be done. Perhaps they should just forget about it and build an outhouse. She called Peter.

"Probably the septic tank is full. I'll have a look next time I'm over."

He left her a note.

"Dear Alice,

I was right about the septic tank. If it's nice next Saturday I'll be over with some tools. You'll all have to help me clean it out.

love, Peter."

Alice drove to the south end to fire him off a postcard —

"Bring a bottle or two of wine. We're going to need it."

On the Friday she baked a special bread from her natural-foods cookbook and made minestrone. She washed her hair, even though she knew she'd have to wash it again the next day.

"We're not carrying shit," the children said.

"It's your shit as well, you'll carry it."

They assembled some old buckets from the shed and two diaper pails which Alice had saved because they might come in useful some day.

Peter arrived about eleven and dug away the earth over the drums. He had borrowed a special drill to get the covers off. Alice was afraid he'd cut himself on the jagged metal.

"It must be teeming with deadly bacteria."

"It's all right."

Both drums were full to the top with stinking brown liquid.

"They're plugged up somewhere, you see. Really it should all just filter out and a slight sludge be left. In England people use it on their gardens sometimes."

"How nice for them. Where are we going to put it?"

"Over beyond the chicken house. In that big hole where I took out that dead tree. Nobody will get into it there and I'll cover it up with earth."

Alice put on rubber gloves and they began.

Bucket after bucket after bucket. It was revolting to think about, and the stench was awful. She wanted to stop and tie a handkerchief over her face, bandit style, but her gloves were covered in filth and it seemed pointless to stop and clean up before they were finished. They all helped except Flora. They wore old worn-out jeans and corduroys from the ragbag and kerchiefs around their heads.

"I hope nobody comes," Anne said. "I'd die."

"It's a pity we aren't horses; horse manure has a nice smell."

"Why is that?"

"They don't eat meat, I guess that's why."

"But chicken shit smells awful and they don't get any meat to eat."

"True. But it doesn't smell as bad as this."

"I wonder if vegetarian shit smells better."

"I have no idea. Why don't you experiment? Suggest it to your teacher."

"Ha ha ha."

In the strangest way, shit-covered and exhausted, Alice was happy. It was because Peter was there and they were all working together. And the fact that he had signed his note "love, Peter." She tried not to think about the fact that he would be gone by evening. She began to sing "In the Quartermaster's Store."

"They have beans, beans, big as submarines — "

The bucket brigade, moving up through the tall grass and across the boardwalks to the dump hole.

''The hens think we're mad.''

''They probably won't lay tomorrow because of all that stink.''

''It must feel funny with a big egg inside of you every day.''

''Weird.''

''What would you do if you got swallowed by an elephant?''

''That's an old one.''

It took them two hours to empty both drums and fill in the dump hole with earth. Then Peter had to do the drainage tiles. All cleaned up and sitting at the kitchen table Alice wondered if maybe he'd stay the night, it was taking so long. The girls had already eaten but that was all right.

She went to the bathroom window and yelled out—

''How are you doing?''

''Nearly done. Will you run a bath for me please?''

The hot water was nearly gone so she heated water in the preserving pan. She emptied that into the tub and laid out a clean towel. She took him out a garbage bag to put his clothes in. A long brown scar ran beside the garden fence. He had made two wooden covers and brought them out from town. They sat on the two drums, now covered and weighted down with flat stones.

''You know,'' she said, ''I think maybe we should get a proper septic tank. Those drums were all right for one old man but that is probably going to happen again.''

''Septic tanks cost a fortune. If it happens again we'll just have to clean it out again. Actually, you and the girls could do it by yourselves now that I've removed the tops.''

He took off his clothes in the utility room and walked naked through to the bathroom.

''Do you want me to wash your hair for you?''

''No. That's all right.'' And shut the door.

Alice walked him down to the car.

''It's been a nice day,'' he said.

"Yes, in a funny way."

At the bottom of the path she put her arms around him and began to weep.

"Oh Peter, you're my oldest friend."

"Hush," he said, patting her hair. "Hush now, I know."

She watched him drive away, Byron running behind the car until they were both out of sight.

In the long June evenings, sitting on the porch, alone or with her children, Alice would hear Stella's car go by on her way to the south end and a visit with Trudl. Or, at one end of the kitchen table, hear Trudl's car pass on her day off. And what she felt was the dark pull of the man (the mandala seemed only an excuse) who was taking her friends away from her, gathering them in under his magician's cape, Peter with his magic box of colors, his warm sympathetic voice ("Alice, sometimes I think I *am* a woman"), his strength.

Will or nil he, he drew these women to him as surely as though he controlled the tide. And she, left behind on the shore, amidst the purple stars, the crabs, the bladder wrack. She might hate them but she could never blame them. The vegetable gardens grew green and full; the air was heavy with flowers, the nights stretched out their long, green hands. Come. Come. Come. Come.

In the long evenings Alice carefully watered and weeded her vegetables and gathered her children to her. All were heavy with the coming summer, drugged with dreams. Sometimes they did not speak to one another (after Flora was in bed), for hours.

Trudl was caretaking a cottage at the south end for a month and working in the café across the road. When the weather grew really warm she and Christobel were going to live in a tent on Stella's father's property. Alice stopped by sometimes but there was distance now between them. Sometimes Trudl would phone and leave a message for Stella to call her back. One day she phoned up and asked if Stella had been around.

"No, she hasn't. Do you want her to call you if she drops by?"

Trudl sounded embarrassed. "Look Alice, Stella's dad's phone is out of order. I have to get in touch with Stella. It's important."

"Why don't you drive up then?"

"I can't. I'm working. Could you possibly go up and tell her to ring me?"

"I can't," Alice said, "*I'm* working. Send a carrier pigeon, why don't you?"

She slammed down the phone.

Although Stella really didn't like to cook, in her shack there were glass jars full of spices and flours and rice, all beautifully labeled in Stella's handwriting. The label for a jar of dried milk was simply a drawing of a breast. Outside near the woodpile the chess set was set up on a stump. Stella fitted right in with her faded jeans and tight red sweater (secondhand of course; she and Harold spent a lot of time in Victoria second-handing). Stella could draw and paint quite well; she had once played the cello. She read a lot and had diaries going back to her teens. She had started a degree when she was down in Berkeley. She had slept with a lot of men — she said so. But kept a small candle burning in her heart for her dead lover, who had asked her to make love to him as he was dying. She couldn't forget him; she had tried but it didn't seem possible. Alice saw Stella's dead lover as the equivalent of Raven's Purple Thing and Selene's new guru in New York. The One Who Knows (or Knew, in Stella's case). Alice did not think she had ever seen Peter in that light, although she had leaned on him, had wanted his steadiness, relied upon it.

Trudl was out of love with Glenn. She needed to be free. She and Stella were very restless. They spent a lot of time together now, practicing their drawing, working out sections of the mandala for Peter to see. Alice felt more and more left out.

One afternoon, flipping through Stella's latest journal (it was always left out on the table) Alice saw once again the drawing of Peter and Stella and Trudl.

"I wish he would leave my friends alone," Alice said. "*I really wish he would leave my friends alone!*"

Stella held out a cup of tea. "I don't want to feel I can't be friends with both of you. And anyway." She smiled her lovely flashy smile. "Someone as special as Peter should be shared."

"I thought you said that you could never share."

Stella stirred her tea. "As a friend."

"Share," Alice said. "Now you sound like one of the hippies. 'I want to share my experience with you, my supper, my old lady etcetera, etcetera, etcetera.' I think you know what I mean. I find it very painful to know that you three are sitting around in my kitchen baring your souls to one another. 'Baring' is the right word, too. There is a little game going on and I don't like it. In fact, it makes me very angry and very sad. There aren't that many interesting people on this island, you know — certainly not here at the north end. You and Trudl, but especially you, have been a great comfort to me — now I feel you both withdrawing. And why? So that you can play silly little games with my ex-husband, whom I still happen to love, as you well know. You have both become the giggly little students and have exalted him to the position of guru. He is a good teacher, I know that, maybe a brilliant one. But you and Trudl aren't *really* doing this mandala thing just to learn about painting, are you? It's a secret society you've formed, with all the attendant whispers and rituals of such a society. We used to have clubs like that in grade school. And they always involved leaving somebody out. Sometimes I think that was their major purpose. You should think about badges and jackets."

When she confronted Peter about the mandala he sighed. "You always wanted to have a finger in every pie."

"That's an interesting image," Alice said, "coming from you."

"You have your writing," Stella said. "We happen to be painters." Alice's mean self felt that Stella would "be" whatever would interest the particular man she was interested in. But again, look what she did. She knew a lot about Peter's likes and dislikes; she left him little gifts of blackberry jam and bottled pears and homemade bread. Once she even left him three poems:

1) I have a hundred foxgloves
in my garden
But no foxes

2) I pricked my finger
I fell down
But no prince came

3) I have five lovers
One of them is much shorter
Than the others.

(She had wondered about doing a drawing for that one — would he get it? Would he think it funny?)

And she left him her daughters, happy (more or less), clean and well-fed, learning new skills and the cottage cheerful with wood fires and candles and usually a soup simmering on the back of the stove.

She put up things on the wall that she thought would amuse him, advertisements from old *Saturday Evening Posts*, drawings by Hannah and Anne and Flora. Left an underlined copy of Durrell where he was sure to see it and pick it up.

She was "being" someone too — someone she hoped Peter would find attractive, would want to get to know again.

One of her friends in town had said, "Nothing is ever final." She hung on to that and tried to believe it.

But now — here was Stella with her Gypsy looks and sexy little nipples (and who had suffered as well, you could see that she'd been through it — you only had to look in her eyes). And Trudl, red-haired like Alice, but younger, soft-voiced, wanting to "go free." And both wanting to learn about themselves, about painting.

And there was Peter, flattered, moved by their stories and their bravery, moved by their mystery.

There was nothing new for him to find out about Alice, nothing. After fourteen years he knew it all.

To get to the lake you drove partway up the Coon Bay road and parked near a stump with three stones on it. That had been Anne's idea, because, at the beginning, when only a few people knew where the lake was, you could go right past. Now that the secret is out — has been out for some years — it is easy to see where cars pull off and in fact, no matter what time of day you go, there is almost always someone sitting on the wet, mossy edge. One of the small resorts even publicizes the lake in its brochure. But that year only a few people knew, Stella and her brother Glenn, Harold, Peter, Alice and the girls. Then Trudl came and eventually told the Coon Bay crowd who arrived with their dogs and their dope and then someone told the resort owners and now a lot of people know.

Someone has built a kind of boardwalk across the huge roots and the mud (in the spring yellow skunk cabbages grow all around — one way or another the lake is always associated in Alice's mind with the color yellow) and across the flat bit, when you come out of the tangled darkness, the smell of mud and skunk cabbage. Now there are Labrador tea bushes on either side. If you crush the leaves between your fingers and then sniff, there is a strong smell of resin or turpentine. The Indians used it for tea and the Coon Bay hippies did too; they said it gave them a buzz. Alice picked some of the leaves and dried them once, made a tea. It had an "effect" but she wasn't quite sure what the effect was. Seven years later the rest of the tea is still in a brown paper bag, in the back of the kitchen cupboard.

Seven years later than that first summer at the lake a lot of things were still in the back of the kitchen cupboard, or in the backs of drawers. Her wedding ring, for instance, which she and Peter had picked out of a jeweler's window in Birmingham. £/10/2/6. Married in a registry office. He was called "bachelor" (holder of a small farm or estate, called in late Latin "*baccalaria*." Remoter origin unknown, and much disputed. Hardly from late Latin "*bacca*," for Latin "*uacca*," a cow). She was a "spinster" (used in Anglo-Saxon as in Du. solely with reference to the feminine gender, but this restricted usage was soon set aside in a great many Middle English words). Afterwards Father and Mother took them out for a nice lunch (the bachelor had a "He-Man Grill," the spinster, in her delicate condition, opted for something lighter, a nice piece of sole).

They parked the car and scrambled down the hill and through the trees to where the path began. Barefoot, and with trousers on or a towel wrapped around to protect their legs. Alice liked the feel of the cool black mud between her toes. Single file through the mud, over slippery roots, holding branches back so they didn't snap in somebody's face (especially Flora's). Then out into the meadow, feet enjoying the warm springy dryness of the moss, the steep cliffs straight ahead. Two blue herons rise at this invasion, fly away protesting. The ground is springier now — they are over a bog, after all. They drop their clothes, towels, glasses, put down the thermos and the snack for Flora, who will spend part of the time sitting on the edge watching and smiling, with her sun hat on, and part of the time in the water. Peter has made her, from Alice's description of one she had as a little girl, a small boat from an old inner tube. There is a seat and a wooden paddle. Flora paddles around while the others swim and at the very end she has a swimming lesson from Harold, who is the only one of them who is really a swimmer, really knows how to breathe properly, except for Peter, who is not of course there but he will be, later on, in August.

There is no shore. The springy sphagnum moss goes right down to the water and the water is deep. Indeed, at the edge, it is a bit like sitting on the end of a diving board, bounce, bounce, bounce, bounce. Alice does not let Flora get too close to this edge and she and the older two always keep one eye on her. Flora is only four and a half. If a bit of the edge broke off she could be in trouble.

The water is cool and dark and their skin really does look yellow from the sulphur or whatever it is in the water. Yellow water lilies too, along the edge. Thick green stalks with fat yellow heads. They look somehow Oriental and slightly out of place. They belong in the ornamental pool of a maharaja, the sacred lingam. Dragonflies too. ''We called them 'darning needles,''' Alice says to Hannah and Anne; they have swum to the end of the lake and back and are now resting on an old tree trunk which has fallen down onto the water, hanging over it, staring down at their yellow legs. The dragonflies startle but they do not sting, put on marvelous aerial displays. Finally they pull themselves out and sit in the warm grass, drying. Someone asks about Raven and Selene. They have gone tree planting up in the interior. They are very much ''together'' again Alice says; they are working things out. There are great bunches of catkins near where they are all sitting. Alice thinks she will come back here in the autumn, pick some and put them in one of the big pots Peter made her one year. Stella is thinking of moving to the west coast of Vancouver Island; she thinks she needs the time alone. She wants to work on a book about her life with her lover that died, the one who asked her to make love to him as he lay dying. She has to deal with all that she says in her rapid, tiny voice, sitting there in the sunlight, pulling on stalks of grass, avoiding Alice's eyes. Harold is going back into deaf society. The girls are going to Peter for a year. Peter is going to live communally, with two other families, in their house in town.

Alice tried not to think about that. With the sun on her back and her daughters right by her, close enough to reach out and touch, she told herself it made a lot of sense to "live in the now." She had had more than thirteen years to make her marriage work and it hadn't. The other women didn't have much to do with it, or the dope. Their ages probably did, and the age in which they lived. Peter would always be more than an aging hippie; he was far too intelligent for that. With someone like Stella (for she knew, they both knew, that Stella was next) perhaps he could combine his role as Peter the Rock with his new sense of himself as Peter Pan. (Or Alice's sense of him. He would say he was "free." She might think of Peter and his band of Lost Boys; it would not be an image her Peter would use.)

In a secondhand bookshop Alice had found, along with a book of Ashanti proverbs, a book of Hausa folklore, customs and proverbs. She liked the way a lot of the stories ended. "The story comes, the story goes. Off with the rat's head." One story was over; she must let it go, let Peter go ("fly away Peter, fly away Paul"), for her daughters' sakes as well as her own. If, deep in her heart, she kept a small hope that he might someday return to her ("come back Peter, come back Paul") she must nevertheless get on with the business of living.

She stood up. "Would anyone like one last dip? The sun will be off the lake in half an hour."

"Flora come," Harold said. He was red all over, or rusty. Alice imagined Pan might have been that color; he went well with the bog and landscape they were in.

"Harold looks as though he was made from this moss and autumn leaves, dry pine needles," Alice said.

Stella smiled. "Isn't he wonderful?" She could afford to be generous now that she was leaving.

"I wonder how he'll make out with his music group?"

"Harold? He'll make out whatever he does. He's one of those people who do."

Flora was swimming on Harold's back, laughing and unafraid. "Let's all go in," Alice said. She didn't want the afternoon to end.

In her physical life Alice was a craven coward but her mind was Mary Marvel, Evil Kneivl, Irish-jawed cops in speeding cars in San Francisco, a trapeze artist without nets, a flying fish on the road to Mandalay, a Hillary always looking for yet another Everest to climb, an Amelia Earhart, a great attempter and achiever. She remembered an elderly grandaunt who'd gone a bit soft in the head. One day, when Alice and her parents were paying a duty visit, sitting on the front porch drinking lemonade and eating buttermilk cookies with scalloped edges, a Saint Bernard walked by.

"My," said Alice's Aunt Aggie, "what a small dog in such a great big box." Alice's mind was the heavyweight champion of the world in the disguise of the Scorpio Housewife. Washing the dishes, she let her mind off the leash, let it run and snuffle, hightail it off the road and into the bushes. Let it bring back prizes. Images, single words. The occasional complete paragraph. Her hands stacked the dishes in the drying rack, opened the fridge to check that there was milk for tomorrow's breakfast, her lips kissed and were kissed good-night. And all the time her mind was making connections, leaping, dancing, Nureyev, new steps, unheard of but graceful positions. Sometimes she wanted to SHUT IT OFF. For her mind was also the broom of the sorcerer's apprentice, the red shoes which kept on dancing dancing dancing, the longest-playing phonograph record in the universe. There were days when she didn't know if she was privileged or cursed. Days when her mind chuntered away like an old man looking out the window:

''Wonder where that little blue boat is goin'. (Pause) You remember in '33, no it must have been '35 because that was the year Clayton had the scarlet fever and the rest of us were sent off to Shinhopple or was it Raquette Lake nobody gets scarlet fever any more why is that? your Aunt Bessie lost her sense of smell where was I a blue sailing boat, oh yes. Well, that same summer was the first time I ever — ''

There were days she wanted to throw a dark cloth over her mind, turn off the lights and tiptoe from the room.

''My mind to me a midden is,'' she once wrote on one of her little pieces of paper.

Alice sat cross-legged high up on the bed in the spare room and talked to Stella while she prepared the supper.

''I like this little room. It's something out of a children's book. And I had a candlestick like this at my grandfather's cottage. It had a picture of Jack-be-nimble on it. Where did you find it?''

''At Saint Vincent de Paul, the last time I was in town.''

''The whole place is nice. We'll have to stop calling it 'the shack.'''

''Well you call yours 'the cabin' and it's not one, is it?''

''No. Just sounds more romantic I guess.''

Stella moved in and out of Alice's line of vision. She thought how well Stella went with the scene. The huge old stove in the middle of the kitchen, the weathered boards, the ancient canning jars with spring tops that she kept her stores in. The ewer and basin underneath the window. It wasn't that Stella looked like a pioneer woman — in fact she was wearing a black Japanese kimono she'd found on her California trip. It was that for some strange reason that was the exactly right thing to be wearing. The black color, Stella's dark hair.

''Perhaps it's the sensuousness,'' Alice said.

''What?''

''Well I was sitting here trying to figure out why you, in your Japanese kimono, look so perfect in this old kitchen.''

"You know what Robert used to say?"

"What?"

"He said I was the 'perfect subject.'"

"You do always look right for where you are. And yet you don't wear the whole counterculture costume. You aren't part of that trip."

"No. I guess I'm just on a little voyage of my own."

"You're glamorous. Unique. Now Trudl's not glamorous. Neither am I."

"No. I see you more as connected with the earth, somehow. Again, not like the hippies or even like Selene, the 'good hippie,' if we can make that distinction. But warm and very alive. It's partly your coloring, like a deer."

"Did you ever see a chubby deer? Trudl's the one like a deer."

"You're not so fat. I was looking at you at the lake the other day."

"I hate my body."

"I think it's kind of nice. Rubenesque."

"Dumboesque. Do you know how Dumbo died? He was hit by a train."

"Oh *Alice*."

They ate in the main room. Stella carefully lifted a pile of drawings off the low table so they could sit under the window and enjoy the sunset.

"Do you know what Christobel did when she was up here the other day? She 'tidied up' the chess game for me when I was busy talking to Trudl. I could have killed her. Glenn and I had been playing for hours."

"I've got to learn to play chess. I tried once, when we were living with Peter's parents. But his father would come up behind me and say, 'Oh I wouldn't do that lovey' and then proceed to play the game for me. So I never learned. But everybody I know seems to play. I like the pieces. It's an elegant game. Much more so than checkers, which is about all I can manage."

"You should learn. It's fun."

"That sensuous thing again. Touching the pieces. I like dominoes for that reason. And even the old checkers that we have, that used to be Peter's father's. So beautifully made. All these plastic games leave me cold."

"Have you ever played Mah-Jongg?"

"No, but I've seen a set. *That* would be a nice game to own. Bamboos. Winds. Dragons. I like the names too."

"I saw a beautiful bamboo-and-ivory set in San Francisco once. But I was on my usual financial tightrope and couldn't afford it. This crab is delicious. And the wine. And thanks for cooking the crabs. I couldn't have done it."

"Hannah and Anne said I was a murderer. I did feel a little funny. I mean, you can bop a fish on the head and put it out of its misery but to cook something live!"

"Have you ever cooked a lobster?"

"Never. I remember my father showing me some in a tank. Some fish shop where he used to go to get something or other. Maybe oysters although I can't imagine that in upper New York State. Anyway, something. And there were all these blacky-green creatures with horrible claws. I was terrified. The fish man picked one out to show me and I ran out the door. They were like something out of a nightmare. Now of course I think they're delicious, but I still wouldn't want to touch a live one."

"I wonder why crabs and lobsters turn colors."

"Perhaps it's just a matter of being scalded to death. You'd turn bright red too."

"Oh. Look at that sky!"

The sun was balanced on the tip of a distant mountain. It was so bright they couldn't bear to look at it and the mountain, because of a natural dip, appeared to be melting where the sun had touched it. The window was open and Alice could smell the sea. Stella had borrowed her father's tape recorder and "The Girl from Ipanema" was playing softly. They were drinking wine out of coronation mugs.

"The west is supposed to be the land of the dead but I love it here. Strange, I'm an easterner, you and Trudl are westerners, I'm an American and yet you two are the ones who've lived in the American west."

"If you can call California the west in that way. Sometimes I think it's a separate country. Invented by Walt Disney."

"I must get there. Maybe if I go to Mexico next year. It will seem funny without you here, or Trudl. She says she's going back to school."

"Yes. I can't imagine going back to school myself, although I suppose I should finish my damn degree sometime. I mean, I'm not qualified to do anything really. A little of this, a little of that."

"You live Stella. Look at the year you've just spent. Look what you've learned about deaf people."

"I've learned that deaf people are pretty goddamned noisy. Well, Harold is. And his friends."

"It must seem very quiet without him around."

"Lovely. Except sometimes I get a little scared at night. Peter Victor came here drunk one night, looking for Harold. Or so he said."

"What did you do?" Alice said.

"I'd locked the door, thank God, and wouldn't let him in, but he could've kicked the door down if he'd wanted to. Or, really, put his fist through the roof, like the big bad wolf. However, he just stood outside laughing for a while and then lurched off. I got the hatchet and put it by my bed."

"Do you think he's attractive?"

"Very," Stella said. "And nice too, when he isn't drunk. I was down at the marina one day to buy some gas and he was on his boat with nothing on but a bathing suit. You should see the muscles in his ass. And he's supposed to be the best fisherman around."

"Would you like to sleep with him?" Alice said.

"Oh, I don't know. It crossed my mind that day but no, not really. I think his wife puts up with enough as it is. She's pregnant again, I see."

"She's so young."

"She won't be young for long at the rate she's going. One little, two little, three little Indians." She smiled and shrugged. "I had a postcard from Harold, if I can find it," Stella said. "I was using it as a bookmark. But which book? Anyway he says he's fine, glad to be back in the deaf society and has lots of girl friends."

"What will you do with this place if he doesn't come back?"

"I don't know. All that work, hey? But I've got to get off this island, I'm going crazy. Maybe Glenn will stay on, although he's so unreliable about his movements he might *not* be a good person to put in charge. I think Harold will come back. He wants to build another boat."

"Do you think he was in love with you Stella?"

"No. Harold has this great love-affair going with himself. And he's quite happy just to move on, see who's around the next corner."

"Like you."

"Oh, I'm not really like that, it's just my mask. I'm looking for a mate."

"I guess I am too. Otherwise it's 'just another body.' I wonder if I'll ever get over Peter."

"Let's have some more wine."

Stella walked Alice across the grass to the road. "I'd walk you home partway but then I'd be scared to walk back."

"That's all right. God. Look at the stars. '*Mica mica parva Stella.*' I remember our Latin teacher saying that the first day. I thought she was so clever. Now I know no self-respecting Roman would write such crap. They're so clear over here, aren't they? I really must learn their names."

The two women stood on the hill in the starry night.

"I'd better go. The kids don't like to be alone too long at night. Glenn said he might drop by, but as you say, he's undependable, and I have to get up early tomorrow morning."

They didn't touch or kiss. Later Alice remembered that they never had. Unlike Trudl and Alice or Alice and Selene. Yet she liked Stella better, found her more real, more honest.

"Have a nice month in town. I'll try and come in one day and we'll go out to the Areopagus for dinner. I hear it's great."

"If you don't get in, don't leave till I come back, will you? I'll be back the morning of the first.

"I'll try not to. I'm pretty restless."

"Good-bye. It was a lovely dinner."

"Good-bye."

Stella turned back toward the shack and Alice went on down the road toward her children. The stars were so bright she decided not to light her lantern.

Neither woman had mentioned the mandala or the fact that Peter, with his portfolio of drawings, the long roll of canvas and the paints, would be driving onto the ferry as Alice stepped off.

And of course Stella was gone from the island when Alice returned.

On the first day of her month in town Alice found a list discarded on the kitchen table:

wineglasses
sleeping bag
acid.

And a note by the telephone: "Everything is in a state of fux."

"I found out today where the word 'ostracism' comes from," Alice said. She said it deliberately, watching her two friends (her strange, estranged friends) across the table at the Areopagus restaurant. Glenn was sitting at the head. They had begun with *meze* and were now proceeding with lamb *souvla* and *pilau*. Trudl hadn't touched the squid. In fact Alice and Stella were the only ones who had enjoyed it. Trudl's tastes were more conservative, in eating anyway.

They were upstairs in the *taverna* and it was getting dark. They had just asked for a candle.

''It's a Greek word,'' Alice said, ''that's what brought it to mind again. 'To banish by a vote written on a potsherd'— a broken bit of pot. Originally a shell; allied to the Greek word for oyster, which was named for its hard shell. Could be written on an oyster shell I suppose and dropped by a sea gull on your head. At first you simply think the sky is falling.''

She took a sip of wine.

''The Athenians banished dangerously powerful or unpopular citizens that way. For as long as ten years. I wonder if they were led outside the city gates or what. This lamb is very good. Are you enjoying the lamb Trudl, Stella, Glenn? I'm so glad we could all get together like this, it's been a long, long time. How's the mandala going? How's the lake? Let's have some more wine, shall we?''

Stella had called Alice and said she was in town and that Trudl was in town as well; also Glenn had showed up at her mother's place and what about that dinner? They decided to try the new Greek restaurant on Broadway. Alice listened to her friend's voice on the telephone and was struck again at what a small, what a *tiny,* voice it was. They were in the same city, just blocks away and yet Alice had to strain to hear her. Surely one's life experiences should be reflected in one's voice? Stella should have Trudl's deep sexy voice, for example. It wasn't fair. Look at all the film stars who were ruined when the Talkies came in.

''Alice,'' Stella said, ''are you still there?''

''What? Yes.'' She looked out the kitchen window at the back garden, which was a mess. The man who had rented the basement suite for the summer was sitting under the apple tree playing his flute. The tree needed pruning. The whole yard was unkempt. They agreed to meet at nine and dine fashionably late. Alice washed her hair but didn't want to join the flute man to whom she had taken an instant dislike for absolutely no reason at all, so she sat on the front steps drying her hair and reading the letters that Durrell had written to Miller and vice versa and wishing she were back on the island with her draft done and a man to love her. One night she had come back, late, to the house in town and found a note taped to the front door.

''Came in to see a movie. Am asleep in the kids' room and will be off early. Hope the work is going well. Love, Peter.''

It was the ''love'' that made her set her alarm and get up to make fresh fruit salad and baking-powder biscuits. They sat at the big table—the novel pushed to one end. The glass door to the small balcony was open and someone was cutting grass. The smell came in along with a hint of breeze. It was going to be very hot, a ''scorcher'' as her father used to say.

''Who's with the kids?'' she said.

''Oh, Stella stayed overnight with them. Alice, why don't you write at your desk?''

''I don't know. I like to spread out. Maybe I got so used to writing at a table all those years that now I can't do anything else. It's a nice desk, however. I'll always treasure it.''

Alice could see he was nervous. After the first fifteen minutes she realized the ''love'' on the note had just been a slip of the pen. He was anxious to be off. What had he come in for? Maybe just a break from the kids. After all, he hadn't been with them steadily in a long, long time. Perhaps he was finding it a strain.

She watched him go off down the street with his small rucksack on his back. She had heard it clink when he picked it up. Wine. For the Mandala Club probably. She had heard nothing from Stella or Trudl all month. Once Peter had phoned to see how she was getting on. He was stoned, she could hear it.

''Oh Alice, the lake, you should see it now. My body turns all yellow. All our bodies turn yellow.'' His thick enchanted voice.

''I know,'' she said, ''I've been there.'' All our bodies turn yellow. After he phoned she was even lonelier than before. The house was too empty. Having a cleanup one day, she found a pile of the desperate letters and poems she had written him last autumn just left on a corner of the bookshelf, weighted down with a stone.

''Once a woman who wouldn't listen was married to a man who wouldn't speak.''

Alice went to see a psychiatrist and cried noisily into her hankie.

''You're unhappy,'' the psychiatrist said in his flat, unemotional voice. Alice saw the tone of his voice like the straight line a dead heart makes on a machine.

''The world's my onion,'' Alice managed, trying to make a joke.

That night she watched *Repulsion*, the ''Four Star Movie,'' and became so scared that she turned it off and then was so frightened at the stillness in the old house that she sat in a chair, afraid to even go to the bathroom, until dawn. It was not the house, of course, it was herself. She knew that but it didn't make things any easier. She sat at tables that Peter had made; she slept in town, as well as on the island, in Peter-built beds. The kitchen chairs were rescued, by Peter, from the university dump. She bought flowers from the Chinese grocers and thrust them into Peter's pots. The great nude, cut off at the head, still lay above the mantelpiece, one leg drawn up, her blue-white thighs locked in a milky dream. She understood why Peter had moved out, even to a dingy basement suite. Maybe they should sell the house, the cabin, split the money and stop being so super-civilized. The trouble was, Peter could fix up another old place or places, she could not. And she had no real job, in society's terms, no credit cards, no credit. How would she keep up a mortgage on the kind of place she'd need to buy without a Peter or a fortune for fixing up? It occurred to her then that perhaps the accepted prison system where men and women were locked into bare, small-windowed cells was not so cruel after all. Perhaps the ultimate cruelty would be to reproduce some beloved room where the prisoner had been happy, and lock him into that. But alone, without the people who made him happy. The shadow without the substance. Everywhere she turned the walls whispered "Peter Peter Peter" or "Peter and Alice" or "our family, our family, our family."

The pigeons on the roof of the house next door called out to her in the evening.

Per — doo — ooo — ooo

"I was actually looking up the root of 'ostentation' when I came across it," she said. "Nowadays of course we don't use oyster shells, we're much less fussy, any old thing will do."

Trudl was wearing the long dress she had bought in California. The candlelight picked up the flame of her hair. The dress was of some thin cotton material and covered in cornflowers. The first time Alice saw it she had thought it was a nightgown. Trudl had laughed. "No, it's my new dress-up dress."

Stella had her hair tied back in a scarf, Gypsy fashion, and had on a black cotton dress from India, decorated with embroidery and mirrors. The mirrors twinkled in the candlelight.

Alice had on a sunset-colored cotton dress from Mexico. She had curled her hair just before it was completely dry and now it hung in pleasant ripples. She knew she was pale but looked good.

Glenn had on his usual plaid shirt. The light bounced off his gold-rimmed spectacles.

They had more wine and drank Spanish coffee. The place was jammed and there was *bouzouki* music playing.

"It makes me want to travel," Stella said.

They talked about travel and books and everything under the sun. But nothing under the moon, oh no, nothing of that at all.

Alice told them that Selene and Raven had come into

town one day and had slept in the backyard ("Of *course,*" Stella said) and that Selene was pregnant.

"I'm really glad for her," Trudl said, "it's what she really wanted."

"*Plus ça change*—" Alice said. "Another little vegetarian for the cause. Actually I'm delighted. She was *glowing,* almost literally."

"And Raven?"

"He appeared to be glowing too. They were on their way up to the Okanagan to pick fruit."

I was silly to come, Alice thought. Trudl and Stella are doing their duty, Glenn is uncomfortable and I'm talking my head off, as usual. Her friends seemed very far away. The whole thing was like a long-distance telephone conversation. Alice thought of an African proverb: "It would be easy to get to Europe if it were not for the sea." She wanted to beg them to come over to her side, only there weren't any sides, were there?

"I want to be friends with *both* of you," Stella had said.

"It sounds good," Alice had replied, "but I'm cynical."

Her napkin started to slip off her lap and she reached down to grab it. Her hand brushed Glenn's leg. He put his hand on her knee and smiled. Oh don't be silly Alice, she thought, he's only a boy. Don't make things any more complicated than they are.

Alice and Trudl and Glenn said they had to go. It was one o'clock already and they were all catching the ferry in the morning. Glenn offered to walk Alice home, it wasn't far, because of the lateness of the hour. Then he'd walk down to his mother's.

The women stood on the corner under the streetlight. They were all a little drunk.

"That was fun. We'll have to do it again."

"See you on the first. If I've finished this draft we'll have a celebration."

Stella and Trudl got into Trudl's car and Alice and Glenn walked away. Glenn put his arm around her shoulders.

"It'll be all right."

"I don't think so," Alice said. "Sometimes, when I'm not loving him, I really hate Peter, I really do. He wants me to stay out of his life but he's always messing around in mine."

He just tightened his pressure on her shoulder and they walked along in silence.

"Look," said Alice, "there's a ring around the moon. I guess we're going to have a storm."

She leaned into him a little as they walked.

Alice had brought the girls into town. She was turning them over to Peter. Peter had been camping for a few days in Garibaldi Park. With Penny. He wasn't there when they arrived and Alice felt strange and lost. Peter's commune was about to begin. Everybody was moving in within the next few days. The bathroom was an Easter purple, the room that had been Peter and Alice's bedroom was egg-yolk yellow with an egg-white ceiling. It was to be a second sitting room so that people could get away from one another. They were using the whole house, upstairs, basement suite, the lot. Something Peter and Alice had never been able to afford to do. She would have to find some place else to stay if she wanted to come in to town. "When can I have a room to write in," she would say to Peter, "when?" But they could never manage it somehow. The children seemed a little lost too and Flora kept grabbing Alice's hand. Alice promised she'd come in for the first day of school. She didn't want to let any of them go and wanted to tell Peter she'd changed her mind. He could get out and his new friends with him, she was staying where she belonged. But had signed a paper, a legal document. Why?

Eventually he showed up.

"Where *were* you?" Anne said.

"Over at Penny's." They liked Penny; his explanation was accepted with no further comment. The phone rang. Peter went to answer it in his new Peter voice. Alice was sitting at the big table in the dining room (the table that Peter had made her from oak — "it will last us forever," he had said) and couldn't help overhearing.

"No" (laughing), "not today. Or maybe later on. I have my 'family' with me." (His voice put quotes around the word.) "Alice brought them in. Yes. No. On the evening ferry I guess, I didn't ask her. What? Half a minute." He came to her.

"It's Trudl on the phone, she'd like to say hello." Alice went into the kitchen — everything is in a state of fux — and picked up the phone.

"Hello there," she said. Trudl asked her how she was.

"The bottom of the lake," Alice said. "Your phrase. How are you?"

Trudl chatted on about the courses she was going to take. She would be in one of Peter's classes. Christobel was enrolled in a day-care center. Life was looking up. Alice thought of Stella's drawing, Peter with an arm around each of her two friends. Anne-Marie had enrolled in one of Peter's classes last year, or was it the year before? Oh yes, last year Penny was up there. The whole thing was getting confusing. Peter was the sun, the hub, around which first Alice, then Anne-Marie, then Penny, now Stella and Trudl, revolved. "Moon ladies," Selene's word. Penny wouldn't last — she was too much like Alice. No mystery, no glamor. Three kids. Too comfortable to be with. Trudl? Stella had said once — they were discussing who would be Peter's next choice — "No, never. I can't exactly say why."

"Not intelligent enough?"

"Perhaps. But that's not really it."

"What then?"

"I don't know."

Alice didn't realize she had put down the phone without saying good-bye. She went in the other room and got her kit bag and her purse. She had some shopping to do.

"Would you like to go out for a coffee," Peter said, "before you leave?"

"No. No thanks. I think we've just about used up all the restaurants in this neighborhood." She said good-bye to her children. They were going to unpack their things and then go to a communal hot-dog roast with Penny and her kids. They were cheering up. She buried her face in Flora's neck. Not even five. Oh God, how would she ever stand it?

Peter walked her to the corner.

"Alice."

"Yes."

"I was traveling with Stella in August. She'll be back on the island to visit her father next week. She wants to see you but we both thought you should know."

"I guess I already know. I probably knew before you did."

Don't step on a crack or you'll break your mother's back.

"You see so clearly Alice."

"Do I?"

"Stella likes you a lot, you know that. She wants to stay your friend. I don't want to come between you and her. You've been very good for one another."

"I'm sorry," Alice said, in her imitation of Harold's voice, a voice from deep inside a cave on the very bottom of the lake. "I can't hear you, I'm deaf."

She signed good-bye to him in Deaf and walked away.

AUGUST

Q. What does a Baby Ghost call his parents?

A. ,,ʎɯɯnW,, pue ,,pɐǝᗡ,,

found in a joke book Anne got for Christmas years ago.

"The real dream was located in an interpersonal zone." Now what exactly did that mean? Since she had opened Stella's present she had been flipping through Cortazar, reacquainting herself. Her dreams had been terrible lately; she had started to write them down.

She dreamed of a shack like Stella's, up a road where there were a few other similar houses. Weathered boards, primitive but beautiful. Her mother came to visit her. Was this only a house where she was staying or was it really her home? Even in the dream she wasn't sure. Alice's mother was impressed. "You see," Alice said, "it's not so bad," or maybe, "you see how beautiful it is." She remembered silvery cedar shingles and the smell of jasmine. Later there were small round tables set for tea. The cups they were drinking out of were made of orangy-red glass. Someone was very angry because nobody had come.

Then she was in a church vestry. Peter was there and she felt very forgiving toward him. The minister took off his surplice and Alice put it on. Then she went into the church, taking off this garment.

The chairs were arranged in a semicircle. There were red, wine-colored hymnal-prayer books. Everyone was singing. Seated ahead of her were two women, one with a peasant's scarf tied over her hair. The child on her knee looked exactly like her mother. Both of the women ahead of her had daughters on their knees. When the mother with the scarf stood up and began to dandle the child Alice could see she wore glasses and was not so pretty or so young as she looked from behind.

A game is played. A child, again a girl, in a very smart two-piece pantsuit is in the center of the floor. They are all singing something about "Wolf, wolf, who has the wolf?" Other children race around, hiding maybe. Finally a man brings the little girl out from behind some curtains where she is crying bitterly. Alice seemed to remember that the child had to kiss a little boy. The little girl didn't look like anyone the dreamer had ever seen before.

And yet the dreamer woke up weeping.

Alice stood naked and alone at the edge of the lake. She stared at the terrible water lilies with their stiff green stalks and swollen yellow heads.

She remembered the laughter, lathering each other's hair, diving down and coming up spouting and laughing, their hair like sea grass floating behind them. Sitting naked in the sunlight talking talking talking, Alice and Trudl and all the children, Alice and Stella alone. The two blue herons. The dragonflies. The smell of Labrador tea.

She shivered and knew this would be her last swim of the season. Remembered a line from *Justine*:

"We have gone different ways. We have all of us taken different paths now."

Remembered Stella and Trudl and herself lying on their backs in the lake, each with an upright water lily in her hand, doing an amateurish variation on a Busby Berkeley routine.

"Okay, slowly to the left now, move." Da da da da. "Kick."

"We should've painted our nipples and our pubic hair green."

"Next time. Let's get the routine down first."

Da da da da da

Then suddenly aware of eyes. Harold/Gabriel at the top of the cliff, stark naked, laughing his head off, clapping his hands.

I have to remember, she thought, all the good times we had. I have to remember that they didn't set out, deliberately, to hurt me. They were just people passing through my life— I'll survive. But she remembered also a sign she had seen in Victoria, "War Amps." (Anne said, "Even their name has been amputated.") Were there love amps too, people who wandered around with parts of themselves, let's take the heart, for example, permanently missing? "Not with a Club, the Heart is broken/Nor with A Stone." Going from door to door selling calendars, key chains, candles. Wearing a badge which echoed the words on the cenotaph. "Is it nothing to you?"

AUGUST

Mrs. Woolf walking into the water with her pockets full of stones. Leonard had said, at one point, that she wasn't strong enough to have children. She leaves a note saying she's been so happy. But her demons were taking over and she was tired of fighting them. Could I have done that—if I'd had no children? That boy in our town—I can't remember his name but his father was a doctor and the boy kept snakes. The mother hung herself in the shower and the whole town was horrified. Mothers didn't do things like that. *My* mother said it was probably all those snakes.

Having children has made me strong—or strong enough not ever to do such a terrible thing. But my demons may be quite tame compared to Virginia's, compared to that woman in the shower. Perhaps I should downgrade my demons to gargoyles, acknowledge that such creatures may exist in the dark corners of my mind but cut them down to size, laugh at them a little. I tend to exaggerate everything—look at how I have mythologized my husband. Some day I must ask myself a very painful question. Did my "love" for Peter, my obsession with him after he left, have anything to do with the attitude of other women toward him? If he was so special that in losing him I felt I had lost my world wouldn't he become exalted in the eyes of others?

If I am convinced that something (or someone) is rare, it's quite possible that others will think so too. Especially if they think *I'm* special, that my judgments are worth something. With Anne-Marie it was probably a question of wanting what I had (or what she thought I had). Wanting it and yet not wanting it — wanting a secret affair. I doubt if she ever wanted to live with any of her "husbands." With Penny I don't know — she may really have wanted Peter simply for himself. She, after all, has not been over here listening to me talk and talk and talk about him. With Stella it probably started because she was bored and restless and because she was curious. Once she said to me that she had a tendency to exaggerate the virtues and accomplishments of her friends. ''Me too?'' I asked. ''Oh no, not you. I have to do the opposite with you.''

Please God let me be able to stay away from that house. Let met not think up silly excuses to go into town, to walk down that street in the darkness just to be close to my sleeping children. Perhaps to stop and stand outside, hoping to hear their voices. There was a man like that once, at the elementary school. Day after day I saw him outside the playground. He always left as soon as the children started coming out. I was worried he was some kind of child molester and so one day I spoke to him. He couldn't stay away from places where he knew his child to be. It was right that the child was with his mother; he didn't want to interfere. He just couldn't stay away. A young man with tears in his eyes. From the smug position of my secure marriage I gave him all sorts of comfortable advice. ''Get ahold of yourself,'' is what I was really saying, ''find some new interests — the well-being of the child is the important thing.''

And they will be out every other weekend. Peter says he needs that, that he counts on me for that. He needs time to be alone with Stella. It's a long drive from Vancouver to the west side of Vancouver Island. And it's only for a year. I have a new calendar which I keep in the bathroom and will take down every other weekend. I have begun crossing off the days.

Section III

''We put out our oars, endeavouring with them to counteract the current, but alas the efforts of the sailors were in vain.''

—*A Spanish Voyage to Vancouver*

In Stella's letter, written on the ferry taking her away: "It will always be THE island, won't it?"

"They've *all* been my friends," Alice said.

"Well," the psychiatrist said, "I'd say he had a neurosis — except that he isn't suffering."

Someone had built a rough boardwalk across the tangle of roots, the ooze of mud, and the lake was now advertised in a resort brochure. The privacy had disappeared but not the mystery. The springy ground they sat on was a thin lid over water that went back — how far? Nobody knew for sure. And where had the other two lakes been, still marked on maps but dried up long ago. What if you went for a walk, pushed your way through the underbrush, explored a little farther? What if suddenly the ground gave way, would you have time to put out your arms or would you be sucked straight down, like water down a waste pipe? People out calling for you, getting anxious as it grows dark and you have not returned. And you, floating just beneath them if they only knew, dead in the yellow-brown water. The lake was so beautiful, a still place, the water cool against your skin, the water lilies with round yellow heads so beautiful against the brown tangle of the bog. The dragonflies, the blue sky, the warmth of the sun. And yet something menacing there too. The sense that the ground might shift at any minute or a huge arm covered in red-brown hair might suddenly shoot up out of the water, grab you, drag you down.

The people who came along from the resort didn't seem to feel any of this. They giggled and squealed, picked water lilies to take home ("Please don't do that," Alice said, "please leave them where they are"), even, occasionally, brought along transistor radios. ("Would you mind turning that off," Alice said, "*please*." Alice saw Flora look at her with a certain embarrassed respect.) Threw matches away in the grass. Eventually she and Flora and Harold and his friend Georgie who was visiting devised a plan. If the people who came walking up to the lake, across the boardwalk, through the meadow of Labrador tea, if these people, say the ones coming now, with their kids and their poodle and their suntan oil, seemed particularly obnoxious — or even just generally obnoxious, they nodded to one another and became a quartet of deaf people, possibly retarded, grunting and signaling to one another, flinging themselves into the water with great splashes, making gestures that could easily be interpreted as obscene. Harold would address the people in his frog-prince voice.

"Snakes," he said, "leeches, big problem." Wagging his wild head and his finger, standing in front of them in all his dripping red-haired glory.

That usually took care of that.

"I wonder what they say when they get back," Alice said. "Let's hope they're too embarrassed to talk about it or the Lodge will figure out it's us."

Harold grinned. It was a bit like a horse grinning, he had such enormous teeth. He loved playing a part, loved showing off for his friend Georgie who had come all the way from

Maryland to visit him. Harold said more deaf people were coming soon. "Deaf society take over island," he said, "people freak out. Wunnerful." Georgie could not speak at all; she only signed and made sounds. But she could write, and carried a pad and pencil with her wherever she went. Flora thought that Harold was in love with her.

"What would happen," Flora said, "if they got married?"

"You mean if they had children?"

"Yes."

"I don't know; I don't know enough about it. Both of them were born deaf; they didn't have meningitis or any disease that made them deaf. So I suppose the chances of their having a deaf child would be pretty high. But I think you're rushing things; they're just having fun."

"Harold says we're the handicapped ones."

"I know. He's always said that. Maybe he's right. I woke up this morning, very early, lay there listening to the birds, thinking how peaceful it all was, and then at about seven somebody up on the ridge started in with a chain saw. That must be one of the most horrible noises in the world. I know the guy probably wants to get his house framed or his wood cut or whatever while the weather holds but it really upsets me. I hate noise."

"But you wouldn't want to be deaf, would you?"

"No. An absolutely silent world, or a world which consisted only of hummings and buzzings, would be awful. Trying to catch conversations, the intense watching one would have to do all the time. No, I wouldn't want to be deaf. Or blind. Being very nearsighted is spooky enough. What I would like is to have all engines and devices like chain saws, jackhammers, portable radios disappear off the face of the earth. People are so afraid of silence. I guess because they might have to listen to themselves. They're afraid of silence and afraid of being alone. It's sad."

"I'm not sure I like being alone," Flora said.

"Well, you are very young. But you like being quiet and you like reading, so you are moving in a good direction. And speaking of the deaf and blind, whatever happened to Nurse Prue?''

''Happily ever after,'' Flora said, ''you were right.''

They read fairy tales. They reread *The Wizard of Oz*, out loud and doing all the voices. ''That was the first movie I ever went to,'' Alice said. ''My mother had to take me out, screaming.''

''Why?'' Flora said.

''I was terrified of the witch's feet sticking out from under the house. I think I was afraid of the Munchkins too. Movies have always had a very powerful effect on me, like being in a dream and knowing you are dreaming at the same time.''

''How old were you when she took you out screaming?''

''Three and a half, four. Something like that. However, I've seen the movie lots of times since then.''

''What was the first movie I ever went to?''

''I don't remember. And I don't remember how old you were when you walked or talked or got your first tooth — I don't remember any of those things and never wrote them down. I have a baby book that was given to me for Hannah — 'Baby's first smile,' 'Baby's first solid food,' that sort of thing, but I never used it. Maybe, because some of those moments were so wonderful, I just naturally thought I would remember. I thought when Anne was born I'd use it, but I didn't, and it's still blank. I'm very sentimental so it surprises me — that I never used it. I'll give it to you for your first child, if you like.''

The night Hannah was born Alice was so keyed up she couldn't sleep. Birth had been so strange, painful and exciting and, at the end, like being scalded, as though all that blood was red-hot, boiling. Hannah was big — the midwives said she looked like a three-month-old. Alice wanted to talk to Peter but he didn't come. He had left at the ending of visiting hours even though she was already in labor. He had been holding her hand but as soon as the head sister rang the little bell out in the hall he dropped her hand and said he must be off. Alice was too proud to beg him to stay. He said "good luck" and kissed her forehead and went away. Then she was really alone and very frightened. A woman was wheeled down the corridor outside Alice's room. She was screaming and shouting—"Oh God, it's comin' oh my God you've got to give me somethin' for the pain." Horrible jagged screams. Scarlet ribbons. And the nurses firm but gentle, talking to that woman as though she were an animal, perhaps a panicky horse, firm, distanced, "Now you must calm down Mrs. Hackett, now this will never do." Nice calm English voices. Voices like firm hands. The delivery room door swung shut on the woman's screams. Frantic Alice pressed the buzzer, over and over again. Finally a nurse looked in. "How are we doing?" she said.

"What's the matter with that woman?"

"Which woman?" the nurse said.

"The one who was screaming. What's the matter with her?"

The nurse laughed. "Nothing's the matter with her, love, she's having a baby." The nurse's teeth stuck out. Alice had been taking natural childbirth classes, reading Grantley Dick Read. It was all a matter of breathing. The nurse examined her—"Oh we've got a long way to go yet."

Sitting up in her flannel nightgown, the thick blood flowing still between her legs, triumphant—the old magic trick, the shell broken, the birdie out, the shell all one again (well, stitched up firmly in places, but all the bits there, all mended), she wanted to talk, to *tell* someone. So she got out her airmail pad and wrote a letter to her child, two hours into this world (you may take one baby step), telling her what it had been like, how it had all been worth it, telling her about the triumph and the terrible exhaustion. Later she tore it up; she was afraid Peter's mother might find it, or Peter himself (for she had also told Hannah how angry she had been, how deserted she had felt when sister rang the visitors' bell and he had dropped her hand, well I must be off). And there would be so many years to go before Hannah would read it.

Alice would have to carry it with her every time they moved. It could easily get lost or fall into the wrong hands.

Years later, walking along the beach, right at the very end of their marriage, walking and talking, Alice carrying a plastic supermarket bag, to pick up any bits of broken glass they came across, Peter said he realized he had ''missed something'' by not being present at any of the births of his children.

He. Had missed something. She wanted to tell him, then, about the fear and anger, about the loneliness. Everyone talked about what a wonderful father he was, how much he helped with the children. She wanted to offer him her memories of those times, like this bag of broken glass. Shove his hand right down into it. Throw it at him. Hard.

''There's a lot of blood,'' Alice said. ''The whole room smells of blood.''

Peter stared politely across to the other shore. He did not like it when she became ''dramatic.''

''Blood everywhere. Blood blood blood blood blood. All over the white sheets. Like an accident in the snow.''

Then it was time to go home.

Now Alice and Flora read fairy tales. They bought crab from the girl who sold fish and they made real mayonnaise. They sat on the front porch sucking red claws and licking their fingers. And then the pale moon rising. Wan. Convalescent. As though exhausted by the sunset. ''It is hard to believe,'' said Alice, ''that all that glory is just dust and atmosphere.''

Some nights they would walk up the narrow road in the dark. Not quite to the end of the island but up the hill and on to the Indian reserve, then cut away to the left and through the woods on a small path. Trees like wrought-iron fantasies all around them. Flora afraid.

''There are so many legitimate things to be afraid of,'' her mother said. ''Cruelty and indifference and wars. You shouldn't use up your fear on the dark.''

''It's not the dark itself,'' Flora said, ''it's what might be in the dark. Waiting.''

"And what? Over here? A raccoon? A deer? There aren't even any bears, you know. And nothing one could legitimately call a serpent."

On nights when there was no moon they carried a strange lantern, "our hippie lantern" Alice called it, made out of a juice can, one end removed, with a candle end sticking up in the middle. The original "handle" had been a piece of cord but later a friend had soldered on a handle made of copper tubing. This lantern cast a soft, round pool of light — all the light they needed — and a breeze could not blow it out. Once they got onto the path Alice would go first and Flora followed, carrying the lantern. They moved past cedars and arbutus quietly, single file, until they came out above the channel. One night Alice said, "It was over there just above that little bay, where Harold and Stella used to live. We went there quite a lot, for visits. Stella rented the place from the Indian band on Kuyper Island. The cabin burned down about three years ago. It wasn't much more than a shack, really."

"I don't remember," Flora said, and Alice wondered, why am I doing this to her?

They crossed some grass and headed for a legless bench set into the rocks. In winter they came here to watch the sea lions fight and play and growl at one another but now, in summer, once it was dark, everything was quiet; all the powerboats had gone, and the sailing boats. Occasionally they could hear a fish boat coming back late, puh-puh-puh-puh-puh; the diesel engine sounded like the panting of a dog.

"I wish I knew the names of the constellations," Alice said. "All those Greek gods and goddesses and mortals tossed up to heaven for punishment or reward." Instead she told Flora about Thoreau's "sky pebbled with stars."

"That's nice," Flora said.

"Do you remember me reading you Wynken, Blynken and Nod when you were little?"

"Sort of."

"There the silvery herring are really the stars. That's a nice image too."

''It's strange,'' Flora said. ''The stars, all that up there, makes me feel really small.''

''You are. We are.'' But it is amazing to me, she wanted to say to her daughter, how large we really are — what an enormous capacity we have for looking, reflecting, inquiring. And how large this moment is for me, sitting here with you, on this still night.

The lighthouse was out of sight, around the point.

''I wonder if dad's sitting out, looking at the stars.''

''He might be. I'm not sure what he's doing this month.''

Peter found a new girl friend about six weeks after he had sat on Alice's couch in her rented apartment. Alice was very much in love and feeling generous so she invited them to dinner. The girl friend looked like a Gypsy, much bigger than Stella, but with the same dark hair. She had a gap between her front teeth and smooth dark skin. She was much younger than Alice, than Stella.

Later Peter had thought he might marry her, have another baby, why not. They were going across Canada by train and then to England to visit his parents. Suddenly, a few weeks ago, it was all off. Flora announced it; she had been there when things had gone wrong. Alice wanted to call Peter up, to say, ''Please come and see me before you go.'' (Oh yes, he was going, as soon as he had his month with Flora, if everything was all right with Alice — and of course it would be — he was so confident he had already booked his ticket, cancelled the original trip for now he was going alone—and flying. Alice wondered if he had taken out cancellation insurance. Now that they were finally divorced, would she still be a legitimate excuse to cancel? ''A little more than kin and less than kind.'')

''It's a pity there are no fireflies here,'' Alice said. ''When I see those stars — on a clear night like this — I often think of fireflies. 'Pale fire,' I guess.''

''Why aren't there?''

''I don't know. That's interesting. I have no idea why. It must have something to do with climate. Anyway, one of the things we used to do, when we were kids, was catch fireflies and put them in jars. Such a strange fire, cold: sometimes it seemed to be pale green, at others yellow. We didn't know then that it all had to do with courtship. We'd sit on the back steps, in the June darkness, with these quart jars in our hands. Then after a while we'd tip them all out again." (Perhaps Flora would not be so afraid of the dark if she had grown up back east. Sledding on winter nights then coming in cheeks burning, feet freezing, for cocoa and baths. Hide and seek in the long slow summer evenings, Kick the Can, sitting on the back steps in the dark, holding jars of fireflies. The dark was their friend—or being outside in it. In bed, in that old house which snapped and cracked like a ship in a storm, was yet another matter. And the yelling downstairs, coming up through the hot-air registers. But even then there was the green eye of the radio dial and her sister down the hall. Had Flora been afraid of the dark before Hannah and Anne went away?)

An orange light approached them from around the point. The night was so silent they could hear, very clearly, the lift and dip of the paddle.

"I think that's Harold," Alice said. "It's so peaceful I don't want to signal to him but I'll bet he's coming to visit. Shall we walk back?"

"He'll be there before us."

"That's all right; he'll wait."

"Good. I want to show him how I've been practicing my signing. Where are the matches?" Flora said, standing up.

"Let's not light the lantern. Our eyes are used to the dark."

Flora hesitated. "Come on," Alice said. "You've got your wonderful brave mother with you."

"Oh sure."

"Flora?"

"All right, let's go."

Two bald eagles were circling high up above the cabin. "Isn't it strange," Alice said to Flora, "how the most magnificent birds have the most awful voices: the eagle, the peacock. And anybody who really wants a man with hawk-like features has never looked at an eagle very closely.''

These eagles had a nest somewhere up behind them and were a familiar sight. Once a bird-watcher friend had brought over a telescope with a tripod and set it up in front of the cabin. The male eagle was up on the ridge, perched high in a tall dead tree. First Flora had a look, then Alice. The bird now looked so close it scared her; it was somehow as though the eagle was on her eye. The sharp beak, nearly as long as his head, the fearsome beauty of it. The talons, to pick her up and carry her off, high above all this. Beak for grasping, for ripping and tearing. Raptors. She stepped back blinking, staggering a little. One wasn't meant to get that near.

''Once, when you were very small, I was sunning myself outside on a blanket while you had your nap. There was a sound, like creaking leather, and a shadow passed over me like some dark angel of death on tattered wings. I shivered, and remembered my mother in the kitchen, thirty years ago, shivering then saying 'Someone just walked over my grave.' ''

They wheeled and wheeled above Flora and Alice, out in the garden picking beans.

''They say an eagle can look at the sun,'' Alice said.

''Why? Why would he want to?''

''That's a good question. Why would he want to indeed?''

''Do you think they are just up there playing,'' Flora said, ''or are they looking for something?''

''It looks as though they are playing but I don't know if birds 'play,' the way kittens and puppies do, for instance. They certainly have lots of courtship rituals—birds I mean. Did you know that when the male hummingbird wants to impress a female, he dives at just the right angle for the sun to catch his throat? Then the female says 'oh wow' or the bird equivalent of 'oh wow' and presumably gets together with him. That's what the peacock's wonderful fan is all about too. But whether then 'play' just to play I don't know.''

''Don't you think it's strange how the male birds and a

lot of male creatures are the fancy ones? Have all the nice feathers and decorations?''

''Well, they do the courting. No doubt that's why.''

''But men and women — ''

''Yes. In our culture anyway. I sometimes think that's what the sixties and early seventies were all about — men had a chance to let loose a little, dress up, wear beads and headbands or velvet coats. Wear an earring. Go without socks. Of course even in our culture men have dressed pretty extravagantly at various times. If we ever go to England I'll take you to the National Portrait Gallery and you can have a look at some of the portraits there. It's a wonderful way to learn about the history of costume.''

''I like dressing up,'' Flora said. ''But girls are lucky. We can wear jeans or skirts. Boys can't.''

''When I went to school,'' Alice said, ''we weren't allowed to wear jeans — we called them 'dungarees' — to school. We had to wear skirts or dresses.''

''*Really?*''

''Really really. I don't know when all that began to change. I think we could wear slacks in the wintertime — it got pretty cold — but no jeans. We had to be 'feminine.' ''

''God,'' Flora said. The colander was full of waxy yellow beans. Alice straightened up, and shielding her eyes, she stared up at the eagles. They had been circling for half an hour at least. What did she and Flora look like to them; were they just part of the landscape, two pale blobs?

''Picking beans is certainly a lot easier than diving for fish,'' Alice said. ''But maybe they enjoy it, the sport of it. Spot the fish, zoom down you go, snap and grasp, up up and away. Sometimes I think I'd like to be a bird. To soar and dip like that. To know precisely what your function in the world was. We pay for our large brain. We have too many options.''

The garden was beautiful now. Ten years of compost and seaweed had paid off. The scarlet runner flowers were open, the nasturtiums climbed the fence; there were dozens of different shades of green. Tomato plants, carrot tops, leeks, beets, swiss chard, lettuces, onions, zucchinis and acorn squash. Rows of parsley and dill. Sage plants, oregano, thyme. This garden was the one place where Alice didn't feel clumsy with her hands, felt she almost knew what it was like to be a visual artist rather that a verbal. Of course she loved the way everything tasted but she liked looking at it too. Sometimes she brought a deck chair into the garden and just sat, perhaps with a sprig of mint or lavender in her hand. Sometimes she fell asleep there, smiling, surrounded by vegetables and flowers, and butterflies and bees.

"Veggies are so wonderful," Selene had written years ago, "because they let you love them." Hippie talk but it was true that you could fuss over them, nourish them, even talk to them if no one was around. They were like babies — you felt protective and proud. But you could also go inside and read your book, leave them out there to get on with it. It wouldn't be a bad place to be buried, she thought now (with those eagles up above, remembering the shadow and sound of wings), underneath this garden, nourishing it. But the kids would think it was ghoulish — "that corpse you buried in your garden/has it begun to sprout." There is an etiquette involved in disposing of the dead. Kittens and puppies in the backyard, maybe, under the apple trees. But mother in the garden, however deeply dug in? And the bones would eventually surface. Mother's index finger, or the whole fist, closed tight around her wedding ring. The dead had to go to designated areas, like immigrants waiting to be processed or people waiting to board planes, dead cheek by dead jowl with a lot of strangers. Maybe that's why cremation appealed. Would they object to mother's ashes? In the garden, probably; but under the Peace roses, maybe not. Who could she ask to do it? None of the girls — they were still too young, even Hannah. Peter maybe — he would understand.

But then they'd never be able to sell the property. It would be a way of asserting her will that she didn't relish. Relish! Ha. "I don't relish leaving your mother there." Already Hannah and Anne rarely came here; only Flora really liked it. For the others, perhaps, there were too many painful memories. Or maybe they had just outgrown it, temporarily. If they had children later on they might want to come back.

''I'll put these inside,'' she said, ''and make some iced tea. Why don't you put the deck chairs in the shade and we'll have a rest?''

Yes, it would be lovely to be buried in the garden, a simple ceremony but some nice hymns from her childhood (no one would know them now; they would have to practice up). ''Once to Every Man and Nation,'' ''Ancient of Days,'' ''Jerusalem.'' And some poems. ''The Windhover,'' ''A Refusal to Mourn. . . .'' She must make a will, but would have to do it some night when Flora was asleep. Or at the last minute, after Flora had gone over to Peter's. Who gets Peter's grandmother's ring? Who gets my great-grandfather's gold watch? Who gets the quilt? And how to dispose of the mortal remains? It was no good being superstitious, wondering if you were tempting fate (''let's take that one, she seems to be ready''); one shouldn't leave all that up to somebody else to decide.

Alice opened the fridge and got out the jug of cold tea. She saw Flora out of the kitchen window, pulling the hose over to the rose bushes, turning it on so the roots could soak. How tan she was getting and her hair was cornsilk now.

''My corn maiden,'' she thought. ''All she needs is a green bathing suit instead of that blue one.'' She decided to take the snapshot camera out as well.

''I am so lucky,'' she thought. ''I just wish I didn't want more, another body in my bed and not just any body, someone who wouldn't mind that I am growing older, that I will always be clumsy with my hands. That I have stretch marks and my tiny waistline is just a memory. That I sometimes snore. I guess,'' she thought, ''I'm talking about a husband, not a lover. I guess I'm talking about Peter if he had stayed.''

''All are provided with some means of securing themselves to the rock, either by root-like 'hold-fasts' in the case of the plants or by a sucker-like foot, strong threads, cement, or claws in the case of various animals.''

Down on the rocks Alice was reading a book about the intertidal creatures. ''Did you know,'' she asked her daughter, ''that the Dungeness crab has a post-mating embrace that may last two days? This book calls them the 'Casanovas of the Beach.'''

''Who was Casanova?'' Flora was lying on her stomach pulling a piece of bull kelp back and forth in the water.

''He was an adventurer who was famous for seducing women.''

''Seems a funny thing to say about a crab.'' She picked up the kelp, shook some of the water off it and put it on her head. The shiny brown ribbons hung down like dreadlocks. ''How do you like my hat?'' she said, ''and how can crabs mate, anyway, with those hard shells on?''

''They discard their shells from time to time and the courtship takes place when the female is discarding her shell for a new one. Very tricky. I wonder how the males know? And I've never seen a crab without its shell. Perhaps the new shell is already there, only very thin and the old one just splits off, like the husk on a horse chestnut. I don't really know very much about crabs. Or about any of these creatures, really. Why the arms will grow back on a starfish, for example, when our arms won't grow back. Why hermit crabs behave in such a ridiculous manner, carrying around houses that are much too small for them, why our friend the crab turns red when we cook him.''

''Or her.''

''I don't think one is supposed to catch the hers. We could go ask the fishermen or the people at the marina.''

''I sure like eating them.''

''Me too. And mussels, And clams. And oysters.''

''Ugh. Oysters. Too slippery and slimy. And they're such a weird color. I only like clam chowder, by the way, I don't like clams by themselves any more than I like oysters. People in Harlequins eat oysters and stuff like that. Not in the nurse books but in some of the others. The rich man takes the girl out to dinner in some fancy place and they eat oysters or clams on the half shell or smoked salmon, things like that.''

''You mean as a first course?''

''Yes. And then they have pheasant, and then he proposes a marriage of convenience.''

''Do you know what that means?''

''It means she'll be the mother of his orphaned child and he'll look after her but they won't fuck.''

How easily she said the word ''fuck.'' That was her fault and Peter's too. They both said it all the time. ''I dropped the fucking iron.'' ''Oh fuck, I forgot my wallet!'' It had no meaning in these statements. She decided to try and reform.

Then, ''That's about it. And of course she's madly in love with him but doesn't know it yet—she will soon—and he's madly in love with her but doesn't want her to get pregnant and suffer and die like his first wife.''

''Do those books ever have unhappy endings?'' Flora said. She had taken off her kelp wig and lain down again, on her back this time, so her tan would be even.

''What do you think?''

''I think not.''

''Bright girl.''

''They always end at the beginning of the 'real' marriage anyway, even if they've been married for ages. They always end when they both admit to one another that they're madly in love. I wonder if I'll ever be madly in love?''

''I expect so.''

Flora sat up. ''Would you ever marry again?'' she said.

''I think so. If somebody came along who could just accept me as I am.''

''You're very nice,'' Flora said. ''You're a bit fat, but you're funny and kind and an awfully good cook.''

''I have nice hair and nice skin,'' Alice said. ''Don't forget that.''

''And a beautiful young daughter.''

''Hmmm. I must take you to see *Lolita.* Anyway, Flora,

I'm not really looking very hard just now. And it does get harder as you get older. Men — men my own age — aren't really interested in me."

"Is it true," said Flora, "that men are more horny when they are young and women when they are older?" And how easily she said "horny." Not a family word. Where did she hear it? At school?

"That's what they say."

"That's weird."

"It makes sure that the young women get impregnated, have their babies young. Or it used to. Maybe a lot of that is cultural as well as biological. Women are having babies later now. They are choosing to do that. For me, I'm glad I had my kids when I was young. It takes a lot of energy. And I was able to work at home, most of the time. That's pretty rare."

"I wonder who I'll marry. Maybe I won't."

"You probably will. But I like to think you won't be pressured into it — by society or by your parents or even your friends. Most of my friends at college were married by the time they were twenty-two. We all did it. You were supposed to be 'on the shelf' if you weren't married by twenty-six or seven."

("I hate marriage," Peter had said, "the institution, that is." Alice had a vision of barred windows, of hands clutching, of desperate faces peering out.)

"Are you coming in again?" Flora said.

"Oh God, it's so cold on this side. You go. I think I'll just sit here and read about crabs."

"Why do we say people are crabby?"

"Something to do with the way crabs move, I guess, or maybe because crabby people often get red in the face? I don't have the dictionary with me."

"For once."

(Alice had called Peter a crab, said he was devious, moved sideways in his relationship with her. He had laughed at that, but nevertheless agreed. He had refused to talk about what was wrong, had skittered away if she tried to comfort him. Went to hide under another stone. And all the time building his new shell, getting ready to back out of the old one and then expand.)

Crabs, she read later, can move in any direction and frequently walk sideways or backwards. Well, that let Peter out. Backwards he didn't walk—wouldn't. He wanted to be her friend. No two-day embrace between them. Or had he been waiting for her to molt and she never did? Had she, in a lot of ways, elected not to grow?

''You are afraid of change,'' he said to her, ''it terrifies you.''

''That might depend on the change.''

(But men my age, thought Alice, wanting to be able to move backwards, wanting to be the Casanovas they never dared to be in youth. Wanting to bathe in pools from which they will rise and rise again, refreshed, rejuvenated. Kissing the firm nipples of flat-bellied women, lapping at cunts as fresh as oysters. Who can blame them?

And what do I want? In which direction, now, do I want to move? Or could I please for the next little while just be a stone, washed by the sea, warmed by the sun, unmoving?)

''At this point they affix themselves to a solid object and proceed to build the shell of the adult form around their body.''

Crustacea, not mollusc, in spite of the way they looked.

''He put her in a limy shell—'' Or she put him.

Oh yes N. Barnacle, God's holy name for us all.

''I can't decide,'' Stella said, ''whether the tables are too high for the chairs or the chairs are too low for the tables.'' Stella and Alice were having a drink in the Bengal Room of the Empress Hotel in Victoria. The waiters were wearing khaki outfits with red-paisley cravats and there was one dried-out tasteless-looking Bengal tiger skin hanging over the fireplace. The little card at their table suggested, on one side, "The Ivory Hunter Cocktail" ("served with ice in a unique ceramic elephant which can be taken home as a reminder of your Safari days in the Bengal Lounge") and, on the other, "Bengal Bites," little snacks to nibble on while you sat across from your ex-friend who now lived in this city with your ex-husband.

"One winter," Alice said, "somebody left us a television set to look after. We watched all the old Tarzan movies on PBS. God they were funny. After a while we turned the sound off. It didn't really make any difference. Always bad white men looking for ivory. Poachers. So how've you been?"

"So-so. I'm working on my degree again but begin to realize why I stopped the last time."

("You stopped because you are essentially a dabbler," Alice said silently. "You will stop again. What you like about the university is that it gives you a chance to be around a lot of men who can teach you things. Guides and mentors." Stella's nipples stood out under her sweater. They always stood out. Alice thought of the Hersheys chocolate kisses her grandfather had sometimes given her as a child. Maybe they got pointy like that—the nipples—from being licked so much.)

The pianist in the corner was a woman. She slipped deftly from "Sentimental Journey" into "My Funny Valentine."

"You should try the Ivory Hunter Cocktail," Alice said. "Or one of us should. I wonder what's in it. I'm just babbling on," she added, "because I can't think of anything to say."

Stella had come to a reading Alice had given at the university. Alice had sent a postcard to Peter, telling him about it, with a postscript suggesting Stella might like to come too. It was Stella who came, alone, just seconds after Alice began to read, and it was Stella who had suggested they go for drinks. They drove downtown in Peter's car (Stella driving, of course) and decided on the Bengal Room because it sounded elegant. It wasn't — or it wasn't now. Like the empire, it might have seen better days. One of Flora's jackets was on the backseat of the car. Alice blinked hard and looked out the window at the pretty gardens. "Somebody called this place 'the velvet-lined rut,'" Stella had said. So all was not total contentment. What if they moved? Did she dare to ask it—are you thinking of moving? No, she did not dare.

"To be up-front about it," Stella said, "I didn't much like being the postscript on Peter's postcard."

Alice shrugged. Should she say, "Well that's what you are now—to me?" Should she say it's interesting how often you use that phrase, "up front." What was the other one? Peter had started using it after he got together with Stella. That's it. "Move your ass." (Move your awhs.) He always picked up the expressions and obsessions of his girl friends. With Anne-Marie it had been ee cummings and mountain mix (the girls, back from a weekend at a cabin on Hollyburn Mountain, talking about mountain mix and Alice asking. Their superior voices. "Don't you know what mountain mix is?") With Stella it was Henry Miller, *The Four Quartets* and "One-eyed Egyptians." The girls again, "Don't you know what one-eyed Egyptians are?" Perfect for Stella, who was always running on about Isis and Osiris and the sacred phallus.

Peter gave the women he fell in love with Rapidograph pens and colored inks and drawings. For the first few months after their separation he had given Alice drawings as well. She kept them all, naked figures of a man and woman almost touching. Always a circular scene—patterns of flowers and leaves—a man and a woman not quite touching, falling forever through space. He pinned them into the rough cedar behind the bed or simply left them on the table after a weekend at the cabin. She saved them all. They were very beautiful and reminded her of Blake, of Botticelli, of a man and a women lying on lush elaborate carpets. She wept when she saw them and yet they gave her hope. They were all together in a plain brown envelope in the bottom of a drawer. She felt, she hoped, they were messages from the real Peter, the one who would eventually come back. Otherwise too cruel. Why send drawings of a man and a woman, naked, the woman's red hair streaming behind her, their hands almost touching, unless he was trying to tell her something? It did not matter that she had seen a similar drawing over at Anne-Marie's, inside the cover of a copy of *October Ferry to Gabriola.* Alice had picked it up because she thought Anne-Marie had borrowed it. Alice had given that book to Peter when they decided they were going to move to the island permanently.

Later, much later, she told Peter how she picked up the book, leafed through it and saw the drawing, of how hurt she had been. That he would give that particular book. It had been so special. Peter's answer hurt even more.

"It wasn't the same book." Anne-Marie now out of the running (if it were a race). They had never, in the end, lived together. And here was dear Stella ("Don't you know about the one-eyed Egyptians mom?") sitting across from her, drinking a glass of dry white wine.

"The origin of the word 'elephant' is unknown," Alice said. "And I guess if we don't stop all those poachers, the creature will be unknown too. I used to love to go out and see the circus train unload outside our town. Once a year getting up very early in the morning and driving out to the flats with my father, the elephants so big and so gentle. Feeding them peanuts and the strange fascination with the ends of their trunks. Like cocks, really, only with a double eye. The way they could bend, move up and down. Of course I didn't see the resemblance then, but those trunks were always very disturbing in an exciting way. How's Peter? Did he send his love? Or his hate or anything at all? A little parcel of concern, something small that you could put in your purse and hand over to me without too much trouble."

Stella lit a cigarette. "He just said he wasn't going to be able to make it." She sighed. "The thing that finally persuaded me, you know, is that he felt no guilt whatsoever. Whatever had been between you was definitely over."

"For him."

"All right. For him."

"Well, I don't think this was such a grand idea after all. I sent that card on a whim. Maybe I thought Peter would come and you would stay away. I always used to read him my new stuff first. Maybe I wanted his praise or his comments or *something*. It shouldn't matter to me what he thinks of my work but it does."

"He admires you very much."

"Ah, admiration. Are you still working at your painting?"

"I do some. I go to a clinic where they do art therapy."

"Art therapy."

"Yes. It's very good."

"Are you not well, then? Are you troubled?"

"You know I've always been troubled. Ever since Robert died. Before. I have a lot of things to work out."

"Why not just do your paintings. Why go to an 'expert' in feeling and interpretation. You're awfully fond of experts."

Stella stubbed out her cigarette and lit another one.

"Is there anything in particular you wanted to talk about?"

"No, no. I thought we'd keep it general. The girls tell me lots of things anyway. You all seem to be one big happy family. I am glad you like them, you know,"

"I love them. They're great."

"Of course you only have the two of them. Three might have been a bit different."

"We see Hannah a lot."

"I know that too. But three is different. And of course they've had all their childhood diseases, things like that. And Flora is old enough to leave with Anne. You do get away by yourselves. With my help too, of course." Why did she keep saying "of course." And her voice sounded so sarcastic. All this would be reported. She's still angry; she's still warped, twisted; she has not yet forgiven.

"I have to catch my bus," Alice said. "I'd better be going now. If I miss this one I'll miss the ferry."

They did not know how to say good-bye.

"Well," Stella said. She didn't get up; her wineglass was still half full (or half empty).

"Well, thanks for coming to the reading. And for playing ladies. See you around and about. Say hello to Peter."

Alice walked away to the strains of "The Girl from Ipanema."

She never talked to Stella alone, again. It was no use; the rage and hurt and sadness were just too much.

And now Stella, in a handwoven shawl, leaning against the castle-hostel which she caretakes with her young lover. "Come dance with me in Ireland?" Stella's father coming up the path to report — they were doing well. Working on the books and articles. And Peter, on Alice's (rented) couch in the (rented) apartment, crying.

On Alice's side of the island there was shore but very little beach. In most places the rocks came right down to the water. But at the boat launch, when the tide was out, there was a stretch of coarse sand where one could sit or dig for clams or turn over small rocks and watch the little purple shore crabs hurry away. Many have legs missing: one, two, three, testimony to narrow escapes from whatever it is that eats them.

Purple very much in evidence here as the starfish, too, are purple, as are the multitudes of mussels which cling to the rocks eternally covered and uncovered by the sea.

The starfish, like the crabs, often have legs missing but they are capable of growing them back again. Not the same legs, of course, new ones. But in the same places. Flora wonders why the starfish can do this and not the crabs, why the starfish can do this and not man. Alice says she doesn't know.

There are hermit crabs, often occupying snail shells or whelk shells into which they can retreat when alarmed. The hairy hermit seems to choose shells that are much too small, like Cinderella's sisters, trying to fit into a slipper they knew was not for them. Alice and the girls have always laughed at the hairy hermit, dragging around his inadequate shelter, imagining he's safe.

Alice goes down to the shore, and to the tide pools among the rocks there, not simply to keep an eye on her child. And anyway, her child does not need that particular kind of care, of keeping an eye on, any more. She goes down because it soothes her and pleases her, all this wealth of life in the intertidal zone. Small fishes, not much bigger than large exclamation points, swim back and forth in the shallow pools. Crabs relax in the warm water. Bits of seaweed sway graceful as hula dancers. Everywhere there are minute snails and chalky barnacles, fixed forever in one spot, condemned to spend their adult lives standing on their heads and kicking food into their mouths with their feet. Alice and Flora — before that Alice and Hannah and Anne and Flora — sit on the rocks or lie, occasionally sticking a divine finger into these small universes, playfully rearranging things. Alice reads Flora "Caliban on Setebos" and she laughs.

"When I was little," Alice says, "my sister and I developed a great passion for killing ants. Just ordinary black ants that we found in the backyard. We would stamp on them carefully, then put their flat black bodies into matchboxes and give them funerals. We had an ant cemetery out near the grapevine, complete with flat stones we used as headstones. It was all very perverse. I don't think we ever told anyone about it and I don't know who thought it up. But now, when I think back on it, I still remember the sense of power I felt. I was so much bigger than the ants. They didn't stand a chance. Now I can't imagine killing anything just for the sheer pleasure of killing, of feeling that power. Not killing because I needed or wanted something for food, not killing because I was threatened, but just killing. And then to give them Christian burials, complete with hymns and headstones!"

"Weird," Flora said.

"Weird is right."

"I've never been to a cemetery," Flora said, "not even to the one at the south end."

"Would you like to go? That's a very pretty one. I often think I'd like to be buried there."

"Dad says he's going to be cremated. Ugh."

"Does he? Well, it's more sensible."

"But to be burned up!"

Alice decided that it would be smart to lighten the conversation.

"I heard a funny story about that."

"*Funny!*"

"Just wait. There was a woman on one of these islands. I forget which one, who always listened to her husband, obeyed him in everything and in every way. He had asked to be cremated and his ashes scattered in Active Pass. So out she went in a boat, accompanied by a friend, and just as she was about to open the package she heard his voice saying, 'Gladys, are you sure you've got the right ones?' She wasn't sure, so she kept the parcel and went back home."

"What'd she do with it?"

"I don't know. That's all I know of the story." Alice laughed. "'Gladys, are you *sure*?'"

They picked a pan of mussels for supper and then the next day they caught a rock cod which had three small crabs in its stomach.

"Does everything just go around eating everything else?"

"Sometimes it seems that way."

AUGUST

Someone must take measurements, decide how wide a sweep the light should have. It will depend on the dangers, of course, on rocks and reefs, even on sunken wrecks. All that has to be taken into account. Guidance and warning are the two purposes of the lighthouse. Come this way, avoid that.

Lighthouses have long been built to conform in structure to their geographical location. Where a good rock foundation existed, as on this island, masonry-tower lighthouses were built, circular in form, with a low center of gravity.

Tallow candles, coal fires, oil lamps were used as illuminating agents. Coal gas, then acetylene. In 1858 electricity was used for the first time, at South Foreland Light, England. Other improvements at this time were the incandescent oil-vapor lamp, rapidly revolving lights, fog bells, whistles, sirens, diaphones and the Fresnel lens, which focuses the beam.

The increasing use of radio beams and radar is making the earlier forms of lighthouse obsolete, with electronics superseding the light signal.

Attendants need no longer live within or adjacent to the tower. Modern lighthouses can function practically unmanned.

Lighthouses date back to the time of ancient Egypt where priests maintained the beacon fires.

Would the wise virgins today be replaced by an electric eye which starts a light when daylight fails? Then if they were out, taking a walk under the stars, bathing beneath the moon, combing each other's hair, whatever, things would still be okay. For who knows when the bridegroom cometh, and who, these days, wants always to be inside, tending the lamps, waiting for his soft knock upon the door?

"Here is another pressed rose," Flora said. "You should have labeled them."

Alice smiled. "I always think I'll remember. What's the book?"

"*The Faerie Queene,* parts one and two."

"Must be my rose, then, but not necessarily. I don't always put them in significant places although it's not a bad idea—then perhaps I'd remember why I'd pressed them in the first place. Use dictionaries and special words. But that might become annoying, roses falling out at you just when you are searching for exactly the right word."

They were having a general cleanup. It had finally rained and Alice woke up feeling that the time had come to set her house in order. Just in case. (Her excuse to Flora being that she wanted it all nice when they came back.) What she really ought to do was throw everything out and start again. Sometimes she felt as though she were living in a museum or one of those places lovingly restored, with a real blacksmith in the smithy, a real farrier in the stables, a real woman, in long dress and white mobcap, sitting spinning or dipping candles. What "real" would she be? Ladies and gentlemen, a real mother.

"Some of these books can go. We'll take them to the thrift store." A lot were books that Peter had left behind. *Be Here Now, Stranger in a Strange Land, Warriors of the Rainbow, Beneath the Wheel*. Textbooks for a course she'd never taken. And the Harlequins could go back to the store. She hadn't realized how many they'd read. *Roman Affair, Living with Adam, Dreamtime at Big Sky, Surgeon at Witheringham, Love and Dr. Forrest* (originally published as *Healing Hands*).

"Who's going to take the Harlequins back," Alice said, "you or me? We've got a couple of dozen."

"You can. I'd be too embarrassed."

"You think I won't be embarrassed?"

"You can say you've been doing research. Going to write a spoof."

"You know, that might be fun to try this winter. Young woman—that's you—comes to exotic island to be housekeeper for lighthouse keeper and his orphaned son. We can steal from John Donne, call it *No Man is an Island*. I can be the bitchy older woman."

"By older," Flora said, leafing through *The Caged Tiger*, "they mean thirty."

''Thank you. Well, we'll subtract a few years from my age and add a few to yours. You can tip over in a canoe and be rescued by the real hero.''

''Who is?''

''I don't know yet. Wealthy owner of a small, private island. We'll figure it out.''

''Perhaps I canoe over to the island to pick blackberries. I don't know it's private.''

''Sounds good, we'll work on it.''

''Why is it,'' Flora said, reading, ''that the original boyfriends always kiss 'clumsily but heartily'?''

''Because they don't know anything about passion. It's passion these girls have to end up with, as well as money.''

Over to the right of the bed, as she stood facing it and joking with her daughter, was a copy of a poster Peter had brought her as a present, the same day he had brought back the gold carpeting for the floor. The Divine Sarah (née Rosine Bernard), dressed up as the Prince of Denmark. Drowned Ophelia horizontal underneath. Suffered an amputation in the year of our lord 1915. Not Catholic, not with Bernard/Bernhardt, so presumably they threw it away, under the bridges of Paris or some such place. Proving what? That you can't make an 'Amlet without breaking leggs? Something like that. You can't have the bow without the wound. ''Numerous farewell tours.'' Alice had been mad on books about legendary women when she was in her teens. Pavlova, Sarah Bernhardt, Florence Nightingale. Now, for the first time, she realized that the slim, boyish-looking woman in the poster resembled Anne-Marie. Was that why he had bought the poster, not as a present for Alice but because it would remind him of his beloved? Oh well, it didn't matter now. Water under the dam. Or under the bridge.

''Knock knock''
''Who's there?''
''Ophelia.''
''Ophelia who?''

Words, words, words, as the prince so aptly said. The beast with two backs. A little more than kin and less than kind. What a lot of damage these mothers do, even when they are queens.

Dressed in a wine-dark costume, the dagger, incredibly, pointing upwards between her thighs. One of Peter's last presents before the split. Perhaps the divine Anne-Marie helped him pick it out. The rat behind the arras.

"And there's pansies, that's for thoughts." A gift of violets came later. And a gift of violence, a swollen, purple eye.

"I want to get rid of that poster," Alice said.

"*Why*? I like it."

"I'm tired of it. I want to throw it away."

"Can I have it for my room, then?"

"I'll still have to look at it every time I go through to the bathroom."

"Oh, come on. If you don't want it, give it to me. You can shut your eyes on the way to the bathroom."

(I stumbled when I saw. Oh yes.)

"Take it then. I think you'll need a knife to get the tacks out, they've been in so long. I'll get a box for all the books."

Next to the poster, directly above Alice's head when she was sleeping, was a mandala done by Selene, "Companionship," based on a quotation by Kahlil Gibran. "Give your hearts, but not into each other's keeping," blah blah blah. She liked the drawing, a multifoliate fantasy flower with the words written around the outside, but the prophet himself was a little too saccharine for her taste. Very popular with the nouvelle vagues. Everything sweetness and light. Everything solved by a maxim or quotation. Everybody Brother Sun and Sister Moon. (But the moon only shining by reflection. Her father trying to explain that to her with a grapefruit, a flashlight and an orange. *Weoxan* and *wanian*. All an optical illusion. Like sunsets, which are only made of particles of dust.)

"For only the hand of Life can contain your hearts." Whatever that was supposed to mean. On the shelf above, two paper fans from Christmas stockings, a postcard reproduction of a pot of Vincent's sunflowers, a ceramic bank from Mexico, half full of pennies (gift from Harold and Stella from a Christmas past), a calabash bowl, a teacup full of burnt-down candle ends. *"Ma chandelle est mor-teh, Je n'ai plus de feu."*

Next to the Selene mandala was another, this one by Peter. A garland of roses and other, more fanciful, flowers, and in the center a man and a woman in his (then) new style, naked, floating, the circle divided like the yin-yang symbol, the woman in darkness with a light body, the man the reverse. If, the next morning, she had put on her coat and taken the ferry across to go and see him? Would it have made any difference?

("I'll do the cookin' honey, I'll pay the rent.")

Hanging from a thumbtacked bit of wool was a little wooden cage containing a toy soldier, one of the grenadier guards. This last a present from Flora, who had tired of the cage once she lost the toy bird who went inside. "He can watch over you mommy." How stupid she was with all her amulets and talismans, her Lares and Penates. A blind man's buffet. Even in the dark she (who was most untidy) knew where each of the special things was. The "snatches of old tunes." Now, like blind kitten or puppy she turned her frantic muzzle toward remembered warmth. A clean sweep was needed. EVERYTHING MUST GO. REDUCED FOR QUICK SALE. Everything out in one swell foop. Tip it all into the sea. Mausoleum this; no wonder she got depressed. Why didn't she have the courage to tell Flora about the poster?

("I'm gonna change my way of livin'
And if that ain't enough
I'm gonna change the way I strut my stuff
'Cause nobody wants ya' when yer old and gray — ")

Learned all those songs at camp. Freak memory. Ask Alice, she always knows the words. Crushes on counselors. Sitting in a circle around the campfire, cups of cocoa in their hands, the only time of day awkward Alice felt at all part of the group. Singing her little heart out. "White Coral Bells,"

''Jacob's Ladder,'' ''On Top of Old Smokey,'' ''The Ash Grove.''

Later, in school and college, more complicated stuff: Plainsong, Vaughan Williams, Bartok. (''Tell me why you come and go daily near my Dwelling.'') All that behind her now. Except for the ones they all sang in the car:

''They have jeans, jeans, as big as submarines
In the corps
In the corps.''

A midden, this place, a midden and a mausoleum. High time she did something about it.

Peter had built deep shelves under the front window. All the games were there: jigsaw puzzles and Lego, Monopoly, Yahtzee, Chinese checkers ''*Jeu de Dames Chinoises,*'' regular checkers ''*Jeu de Dames,*'' ''Ladies' game.'' Because it was simpler than chess, perhaps? But ladies were not simple and it wasn't a game, it isn't a game; it only seemed that way sometimes. That African proverb: ''By the time the fool has learned the game, the players have dispersed.''

How to prepare Flora? Not possible. She'll go where her heart leads. She may be hurt, but there is no way to stop that. Just try and make her strong. There are no rules. Each time it will be different. Will it, perhaps, be a little *easier* for her generation? Not so many games?

Dominoes, a felt-lined wooden box full of chess men which she really should send over to Peter. Card games of all sorts, *Milles Bornes*, Rook, Happy Families. Did happy families, like happy women, have no histories? Is that what Tolstoy meant? A miniature farm in an old biscuit tin. A spirograph for making fancy patterns. A bead loom with something started on it and never finished. Next to all that the collection of *National Geographics*. ''The Crab that Shakes Hands,'' ''Keeping House in a Cappadocian Cave,'' ''Childhood Summer on the Maine Coast,'' ''Where Elephants Have Right of Way.'' (''To British Photographer George Rodger and his wife Jinx, Africa has long been a second home.'') Never photographs of the authors' wives, bare-breasted. Never disease or dirt. No yaws, leprosy, elephantiasis. A curious magazine, a wedding gift from Alice's grandfather and then kept up by them after he died. They all read it—or looked at the pictures anyway, cut them out for social-studies projects. Armchair adventurers. But she supposed they'd better stay. The tattered collection of *Saturday Evening Posts*. Alice's manuscripts wrapped up tight in green garbage bags and labeled "first draft," "second draft," "final." All of it better stay. But could at least be taken out and dusted.

They began on the steamer trunks. You could still read the faded labels. Destination LIVERPOOL, SOUTHAMPTON, MONTREAL. One of them said, in bold letters, NOT WANTED ON VOYAGE. How many times had she gone back and forth across the water. First as shy young virgin and then. And then. Peter takes Alice's picture in front of Nelson's Column. They have come down to London to go to the immigration office with their baby. She is three months old and they show her Buckingham Palace. "The King said he was sorry/ so did the Queen and Prince/ James James Morrison's mother hasn't been heard of since." D deck on the old *Empress of France*. Good-bye, good-bye, good-bye. Alice took the baby down below; it had started to drizzle. Peter's parents diminished to a manageable size as the ship pulled away.

"The Last of England."

"Were these *mine*?" Flora said in astonishment, holding up a tiny pair of white-kid shoes, hardly bigger than shoes for a rather large doll.

"Mine," Alice said. "Grannie gave them to me the last time I went to see her."

Flora held them in the palm of her hand.

"How strange that we can't remember any of that."

"Very strange. All we know is what our parents and relatives choose to tell us."

''And a lot of that is probably lies.''

''I'd say 'myths,' not lies. Not usually.''

''I suppose you want to save these,'' Flora said.

''Need you ask?''

''The giveaway pile isn't very large.''

''No. I find it hard.''

Flora was much more tidy than Alice. She was beginning to wear a frown of disapproval. Should Alice simply hand the whole thing over to her daughter? Let her be ruthless? No. It was up to her.

But look, here at the bottom of the second trunk is a teal-blue strapless taffeta evening dress.

''Far out!'' Flora says, holding it against her. ''Can I have that?''

''It won't fit you yet, but soon. Not in style, however. Too stiff and fancy. I doubt you'd ever wear it.''

''How old were you when you wore it?''

''Sixteen. I didn't have very big boobs so I stuffed my strapless bra with old nylon stockings. Spent the whole night worrying about whether the boy would know or not.''

''Did he?''

''If he did, he didn't let on.''

''Let me have it. I can use it for fancy dress even if styles don't change.''

''It's yours.''

Flora went to put it away in her bottom drawer.

''Flora,'' Alice called, ''I think it is time we stopped for tea.'' The rain hadn't let up. She could almost hear the dry earth sucking it in.

AUGUST

''We immediately realized the danger which we should

be in among these islands, the channels between which we did not know and which we had no interest in exploring."

On the last morning Alice put Flora on the ferry. Harold and his friend Georgie were going over on the same ferry so Flora would be fine. Flora hugged her mother and said, "Have a good time mum," and then, remembering, said, "oh" and looked embarrassed.

"It's all right," Alice said. "I'll let you know right away if there are any handsome doctors."

She stood waving, watching the ferry, until it disappeared. Then she went home, dug in the garden, scrubbed the kitchen floor, went for a walk, kept herself busy until the dark. Then sat on the porch having a final glass of wine. Byron and the cat lay at her feet. What if she said to the surgeon, "And while you're at it, would you remove my memory?" Or remove the sad parts, anyway. If they could inject a dye so that the sad parts would show up one color (blue, of course, indigo dark blue) and the happy ones another. (Yellow as daffodils or grass green.).

"I'm a good woman," she thought. "I let myself be persuaded that I wasn't, that I was, in my personal relations anyway, a failure. And, like a rider on a teeter-totter, the farther down I went the higher up I pushed the rest of them: Peter, Selene, Trudl, Stella. Not really aware, until this summer, that it was I who was doing the elevating. The arrival of Harold, plus what's going to happen or not happen the day after tomorrow, has certainly brought it all back. Imagine old Harold being the equivalent of Proust's teacake." She scratched the cat behind her ears. "I forgive," she said, "but I don't forget. And that's hard."

"Can't get over it."

"Can't get over it."

"Gotta go through it."

''Gotta go through it.''
''All right.''
''All right.''
''Let's go.''
''Let's go.''
She blew out the candle and went inside.

Put out the cat. Wind up the clock. Nice to think of it still ticking away tomorrow.

''Wash your hands, put on your nightgown; look not so pale.''

Everything ready for the morning — shipshape. If life were a rose, you could press it:

I'm afraid.

She just didn't care. Floating. Floating. But not facedown, not the dead man's float. Or not yet, ha ha. Floating down the hall, horizontal woman. On a trolley, covered by a white sheet. Something ordered from room service à la carte. Horizontal in the big elevator, descending. Somehow she had thought the operating room would be up. Dropping down, it seemed a long way. Any minute now the White Rabbit would appear. No — he'd be colored. Pink. Blue. Primrose yellow like her room. All cheerful colors now. White banished. This ain't a hospital it's an hotel. This is not a scalpel in my hand, it's a dinner knife. The surgeon almost as handsome as the silverware.

In the anteroom, the anesthetist and his assistant wore green. She could not tell, from the bulge where his nose was, whether he had hawklike features. Anyway, he was too old. She could see wisps of white hair beneath his cap. Silver at the temples, yes; white hair no. The anesthetist's assistant had a round rosy face and wore a red-calico kerchief over her hair.

''You look like something out of the *National Geographic,*''

Alice said, "The Hutterites, Plain People of the West,' 'Following the Reindeer with Norway's Lapps.' Something like that." The girl was not masked and had a pretty smile.

The doctor put a red-rubber tourniquet around Alice's arm, sponged it with alcohol.

"I can't do this," Alice said, suddenly wide awake. It wasn't funny anymore. She wanted to see Flora one more time, Anne, Hannah. She wanted to see Peter. There were so many things left unsaid. She had never made her will.

"Just relax," the anesthetist said, "take a deep breath."

Alice turned her head away. The Hutterite smiled and took her hand.

"That's the idea," the doctor said, "very good."

The needle went in.

"Now will you just count for me, backwards and from ten?"

Alice in a last burst of naughtiness took her good hand—the one not strapped to a board—and began signing.

"10"

"9"

"8"

Her hand fell back on the trolley. "All right," the doctor said, very businesslike. "She's under now. Let's go."

Flora and Peter were out in the red rowboat, fishing.

"Do you know what time it is?" Flora asked.

Peter looked up at the sky, squinting.

"I don't have my watch but I'd say it's getting on for noon. Are you hungry? Shall we turn back?"

She didn't answer and he saw that she was crying; huge silent tears were rolling down her cheeks.

"It's going to be all right, you know," he said. And

then, because he wasn't quite sure how to comfort her, he said,

''Flora, would you care to row?''

She nodded silently and, bracing their fishpoles beneath the seats, they carefully changed places.